Automotive Oscilloscopes

Preface

Welcome to Automotive Oscilloscopes

This book has been written because the rapid growth of technology used in cars has highlighted the need for a thorough approach to vehicle diagnosis and repair.

An OBD scan tool is vital for modern vehicle diagnostics; however, trouble codes will only take you so far. The problem can arise when the phrase 'fault code' is used in connection with diagnosis. A code will rarely point you directly to the root cause of a vehicle fault, but can help focus your diagnosis on a specific area and run functional tests etc.

It is the oscilloscope (or scope) that can truly test the operation and health of a system component.

An effective diagnostic routine should always begin with a logical assessment of symptoms, and then uses reasoning to reduce the possible number of options, before following a systematic approach to finding and fixing the root cause.

A very important thing to remember about oscilloscopes, is that they should be easy to set-up and use; otherwise they'll be passed over for a more familiar tool within your comfort zone.

Remember that nothing ever happens within your comfort zone.

There is a great deal of misconception about how difficult a scope can be to set-up, and once you are used to your own equipment, if it is laid out and ready to use, it will soon become your diagnostic tool of choice.

This book has been written to help you get the most from your oscilloscope, and has been designed to give straightforward and uncomplicated methods that can be used effectively for automotive diagnosis. Further in-depth training can then be undertaken to help reinforce the concepts explained.

The chapters will introduce you to health and safety, electrical principles and the set-up and use of oscilloscopes, including key terms, points of interest and diagnostic tips to support the information provided within the text.

Chapters:

How to use this book…………………….……**Page 3**	
Oscilloscope quick set-up guide………….**Page 6**	
Chapter 1 Electrical Fundamentals………...**Page 8**	
Chapter 2 Introduction to Oscilloscopes……………………………………**Page 20**	
Chapter 3 Automotive Actuators and Waveform Analysis……………………………………………...**Page 33**	
Chapter 4 Automotive Sensors and Waveform Analysis………………………………………..….**Page 83**	
Chapter 5 Automotive Ignition Systems and Waveform Analysis……………………..…….**Page 131**	
Chapter 6 Automotive Network Systems and Waveform Analysis……………………..…….**Page 150**	
Chapter 7 Automotive Pressure Testing and Waveform Analysis……………………..…….**Page 170**	

This book offers:

Information to help Automotive Technicians systematically diagnose electrical and electronic vehicle faults using an oscilloscope.

Ideal support for learners and tutors undertaking automotive qualifications.

A large number of illustrations to support knowledge and understanding, with an analysis of automotive waveforms.

Automotive Oscilloscopes

Text © Graham Stoakes 2017

Original illustrations © Graham Stoakes 2017

The rights of Graham Stoakes to be identified as author of this work have been asserted by them in accordance with the Copyright, Designs and Patents Act 1988.

Cover design - fiverr.com/alerrandre

Published by Graham Stoakes

First published 2017

First edition

ISBN 978-0-9929492-6-6

Copyright notice ©

All rights reserved. No part of this publication may be reproduced in any form or by any means (including photocopying or storing it in any medium by electronic means and whether or not transiently or incidentally to some other use of this publication) without the written permission of the copyright owner, except in accordance with the provisions of the Copyright, Designs and Patents Act 1988 or under the terms of a licence issued by the Copyright Licensing Agency, Saffron House, 6 - 10 Kirby Street, London EC1N 8TS (www.cla.co.uk). Applications for the copyright owners' written permission should be addressed to the author.

Acknowledgements

Graham Stoakes would like to thank Anita and Holly Stoakes for their support during this project.

Thank you to alerrandre for the cover design.

The author and publisher would also like to thank the following individuals and organisations for permission to reproduce photographs:

Simon Fred Kinchin 22

Cover image: Shutterstock.com - adike

Author

Graham Stoakes AAE MIMI QTLS is a lecturer and author of college textbooks in automotive engineering for light vehicles and motorcycles.

With his background as a qualified Master Technician, senior automotive manager and specialist diagnostic trainer, he brings over 30 years of technical industry experience to this title.

www.grahamstoakes.com

Some of the waveforms and analysis contained in this book are based around the automotive tests included with PicoScope, the most popular Automotive Oscilloscope.
More information and some really useful additional resources can be found on the Pico Technology website www.picoauto.com.

Introduction

How to use this book

Could you spot if someone tried to badly fake your signature?

Are the curves, in the wrong place?

Do letters slope upwards instead of downwards?

Are there letters missing completely?

If so, then you can use an oscilloscope for automotive electronic diagnosis.

Every electrical and electronic component found on a car has its own signature, and once you can identify the forgery, you're on your way to discovering the problem. The oscilloscope allows you to have a viewing window on the normally hidden worlds of current, voltage and pressure. Once you have become familiar with the function and operation of automotive electrical circuits, and what the components do, you'll be able to analyse a representation of their signatures and spot the fake.

This book has been fully illustrated to help give you information about the waveforms produced and procedures used when conducting diagnosis with an oscilloscope. Unlike many publications or training manuals, it has been written with an aim to help technicians become confident in the use and set-up of automotive scopes.

Due to the wide range of circuits, components and operating faults possible on a vehicle, this publication tries to remain generic, allowing you the flexibility to adapt the content to your diagnostic needs without tying you down to specific settings or figures. It concentrates more on the shape of the waveforms and what different parts mean.

Throughout this book, you will find features that aim to support and enhance your understanding and use, including:

Information in these boxes will help indicate safety features that you should consider when conducting work on vehicles and electrical circuits.
They are designed to reduce the possibility of injury, or damage to vehicles or equipment.
Even if no specific safety advice is given, you must always assess the potential risks in any activity or diagnostic routine.

Information provided in these boxes is designed to support the use of oscilloscopes in automotive diagnostics.
They give material which will help understanding and reinforce knowledge on systems, components and testing.

Information in these boxes describe key terms related to the subject of automotive diagnostic testing.
If technical vocabulary is understood and used in the correct context, it provides a basis for good practice when undertaking repairs.
Within the text, words highlighted in **bold** will have a definition described by this feature.

Introduction

Information in these boxes provide handy diagnostic tips when working on specific systems or components.
Although they may not all be relevant to the vehicle or task you are conducting at the time, they will help give you ideas that you can adapt and use within your systematic diagnostic routines.

Electronic and electrical safety procedures

Working with any electrical system has its hazards and you must take safety seriously. When you are working with light vehicle electrical and electronic systems, the main hazard is the possible risk of electric shock. Although most systems operate with low voltages of around 12V, an accidental electrical discharge caused by incorrect circuit connection can be enough to cause severe burns. Where possible, isolate electrical systems before repairing or replacing components.

If working on hybrid or fully electric vehicles, take care not to disturb the high voltage system. The high voltage system can normally be identified by its reinforced insulation and shielding, which is often coloured bright orange. These systems carry voltages that can cause severe injury or death.

Always use the correct tools and equipment. Damage to components, tools or personal injury could occur if the wrong tool is used or misused. Check tools and equipment before each use.

If you are using electrical measuring equipment, you should check that it is accurate and calibrated before you take any readings.

If you need to replace any electrical or electronic components, always check that the quality meets the original equipment manufacturer (OEM) specifications. (If the vehicle is under warranty, inferior parts or deliberate modification might make the warranty invalid. Also, if parts of an inferior quality are fitted, it might affect vehicle performance and safety). You should only carry out the replacement of electrical components if the parts comply with the legal requirements for road use and environmental protection.

Although oscilloscopes can be used for testing the high voltage systems of hybrid and electric vehicles, you <u>should not</u> attempt diagnosis and repair of these vehicle types unless you have had specific training.

Provision and Use of Work Equipment Regulations 1998 (PUWER)

The equipment used in your workshop needs to be:

- Safe to use.
- Maintained correctly.
- Inspected regularly.
- Only used by people who have received appropriate training.

The Provision and Use of Work Equipment Regulations 1998 (PUWER) place the responsibility for the safety of workplace equipment on anyone who has control over the use of work equipment, including your employer, you and your colleagues.

Introduction

Personal Protective Equipment (PPE) at Work Regulations 1992

This regulation requires that employers provide appropriate personal protective clothing and equipment for their employees. It is your duty to use PPE if required.

  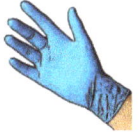  

When selecting PPE, make sure that the equipment:

- Is the right PPE for the job – ask for advice if you are not sure.
- Fits correctly – it needs to be adjustable so it fits you properly.
- Is properly looked after.
- Prevents or controls the risk for the job you are doing.
- Does not create a new risk, e.g. Overheating.
- Is comfortable enough to wear for the length of time you need it.
- Does not impair your sight, communication or movement.
- Is compatible with other PPE worn.
- Does not interfere with the job you are doing.

Vehicle Protective Equipment (VPE)

To reduce the possibility of damage to the car, always use the appropriate vehicle protection equipment (VPE):

 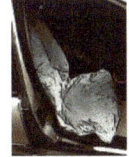  

Wing covers   Seat covers   Steering wheel covers   Floor mats

Quick Set-up

Quick set-up guide

An oscilloscope is a piece of electrical test equipment designed to act like a voltmeter or an ammeter.
A multimeters' measurement display can't change fast enough to deal with modern electronic systems on motor vehicles – the numbers on the screen can't keep up. The answer to this is to use an oscilloscope.
Unlike a voltmeter, oscilloscopes not only show volts or amps but also time.
Instead of a digital readout, the results are shown as a graph of volts or amps against time on a screen (as shown in Figure 0.1).

- A point on the display screen shows the measurement taken from the circuit and then moves across the screen left to right. It leaves behind it a trace, showing a history of its journey and this is known as the **waveform**. When the trace reaches the right-hand side of the display, it resets to the left-hand side and starts over again. (This is known as **triggering**).
- The graph normally shows voltage or amperage at the side of the screen (on the Y-axis) – this axis is often called **amplitude**. Use the scale setting switch in a similar way to the dial on a manual multimeter to choose the amount of volts or amps that are shown on the screen.
- The graph normally shows time across the bottom of the screen (on the X-axis). This axis is often called **frequency** or sweep. Use the timescale switch in a similar way to the dial that is used to choose the amount of volts on a multimeter.

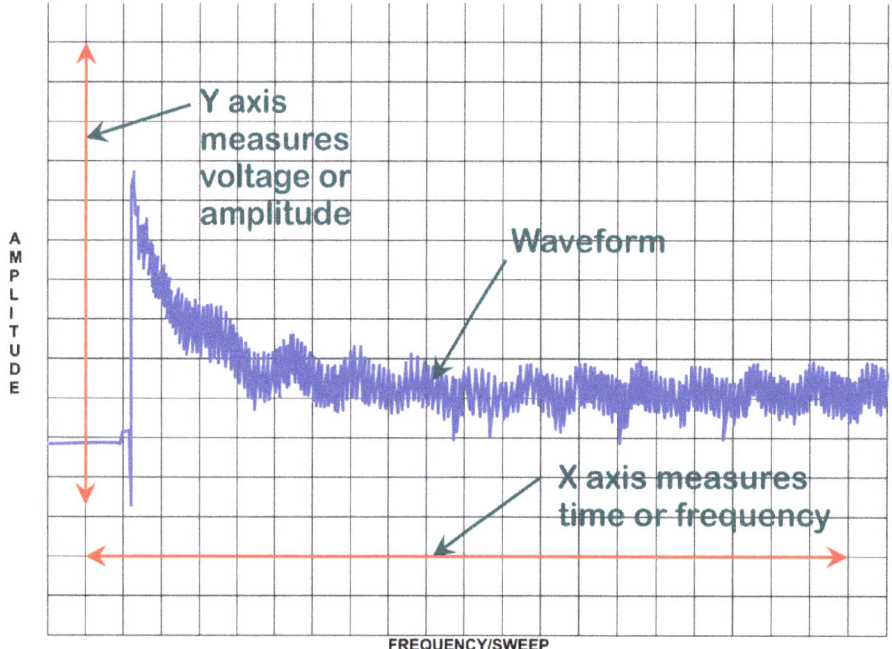

Figure 0.1 An Oscilloscope screen

An easy way to remember which axis is which on a graph is to say 'X is across' (a cross).

Quick Set-up

If you are unsure of the most appropriate voltage or timescale to use for a particular test, simply select a value somewhere in the middle and then move up and down the scales until the pattern is displayed to your satisfaction.

Lots of people are put off using oscilloscopes by the large box containing many wires and connectors. They feel that it will be complicated and time consuming to set up, so they don't bother.

However, to use an oscilloscope for simple electrical testing, you only need two probes – a common and voltage wire – just like a multimeter.

To measure amperage, you may need an inductive clamp and HT ignition systems will require a secondary probe. Most of the diagnostic sockets found on oscilloscopes are colour-coded, so after a quick check of the manufacturer's instructions, it should be fairly easy to know where to plug these probes in.

Waveform – the line traced on the screen of an oscilloscope as measurements are taken.

Amplitude – the height of a waveform, measured in volts or amps.

Frequency – the time scale of a waveform (how often something happens).

Triggering – the point on an oscilloscope display when the waveform refreshes and starts again.

4 step guide to connecting an oscilloscope

How to:

Note: The oscilloscope probes may come in different colours, but for the sake of simplicity we will call them red and black here.

Step 1
- Connect the tip of the black lead to a good source of earth, such as the battery terminal, metal bodywork or engine. This will then only leave you with the red wire to worry about.

Step 2
- Now connect the red probe to the circuit to be tested.

Step 3
- Adjust the scales until you see an image on the screen.

Step 4
- After some practice, you will become familiar with the patterns and waveforms created by different vehicle systems.

Chapter 1 Electrical Fundamentals

This chapter will help you develop an understanding of fundamental electrical principles used in automotive engineering. It also introduces the basic operating theories of electricity and electrical systems that will aid you when undertaking maintenance and repairs. Remember to work safely at all times and observe the relevant health and safety regulations; while developing diagnostic routines that are systematic and effective.

Contents
Electrical and electronic units 8

What is electricity? 9

Series and parallel circuits 13

Ohm's and power law 13

Common electrical faults 15

DIN terminal numbers and breakout boxes 18

Electronic terminology 18

Electrical and electronic units

In cars, electrical energy is created by a chemical reaction (in a battery for example) or by the disruption of magnetic fields near electrical conductors (in an alternator for example). A description of the main electrical units is shown in Table 1.1.

Table 1.1 Electrical units

Volts 	Voltage is electrical pressure. Voltage is the potential force in any part of an electrical circuit, and is named after Alessandro Volta. Two main types of voltage occur in electrical circuits: **Electromotive force (EMF)** is potential pressure, and is usually considered to be the open circuit voltage when all electrical consumers are switched off and no current is flowing. **Potential difference (Pd)** is the voltage drop caused by flowing electricity when the circuit is switched on.
Amps 	Amps are the units used to measure the amount of electricity in any part of an electrical circuit and are named after André-Marie Ampère. Amps is measured when electricity is allowed to flow in an electrical circuit – this is known as **current**. There are two main types of electrical current: Direct current (DC) is electricity that flows in one direction only. Alternating current (AC) is electricity that moves backwards and forwards in an electric circuit. Amperage is the same wherever you measure it in the circuit (at the beginning, in the middle or at the end).

Electrical Fundamentals

Ohms

Ohms are the units used to measure the resistance to electrical flow and are named after Georg Ohm.

Resistance has a direct effect on the operation of any electrical circuit as it tries to slow down the flow of electricity.

As resistance rises in a circuit, current and voltage fall, which can restrict the operation of electrical components. In some electrical circuits, resistance can be used as a method of control for electrical components, but in most circumstances a high resistance is undesirable.

Watts

Watts are the units used to measure electrical power made or consumed and are named after James Watt.

Power is defined as the rate at which work is done. When referring to electrical components, the higher the wattage, the more powerful the component will be and the more electrical energy it will use.

Electromotive force (EMF) – the open circuit voltage when everything is switched off.

Potential difference (Pd) – the voltage drop in an electric circuit when it is switched on.

Current – flowing or moving electricity, measured in amps.

What is electricity?

Every substance known to man is made of molecules. The molecules of a substance are made up from atoms. For example, if the substance is water, the molecule is H2O. This means that the molecule is made up of two hydrogen (H) atoms joined to one oxygen (O) atom.

Figure 1.2 The atoms in water

Electrical Fundamentals

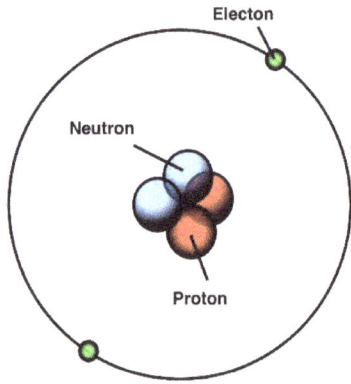

The reason why it can be difficult to understand electricity is because it is contained within atoms. Atoms are very small and hard to imagine.
The easiest way to imagine an atom is like a miniature solar system, with a sun in the middle and planets orbiting around the outside.
In the case of an atom, the nucleus represents the sun. The nucleus is made of positively charged particles known as protons. It also contains particles with no charge known as neutrons.
Orbiting around this nucleus (in a similar way to the planets) are negatively charged particles known as electrons. As the name suggests, it is the electrons that produce electric current.

Figure 1.3 A helium atom

Different atoms have different numbers of protons and electrons, as shown in the periodic table – a chart with the known elements laid out in order of atomic weight (or the number of protons in their nucleus.

1 H																	2 He
3 Li	4 Be											5 B	6 C	7 N	8 O	9 F	10 Ne
11 Na	12 Mg											13 Al	14 Si	15 P	16 S	17 Cl	18 Ar
19 K	20 Ca	21 Sc	22 Ti	23 V	24 Cr	25 Mn	26 Fe	27 Co	28 Ni	29 Cu	30 Zn	31 Ga	132 Ge	33 As	34 Se	35 Br	36 Kr
37 Rb	38 Sr	39 Y	40 Zr	41 Nb	42 Mo	43 Tc	44 Ru	45 Rh	46 Pd	47 Ag	48 Cd	49 In	50 Sn	51 Sb	52 Te	53 I	54 Xe
55 Cs	56 Ba	57-71	72 Hf	73 Ta	74 W	75 Re	76 Os	77 Ir	78 Pt	79 Au	80 Hg	81 Tl	82 Pb	83 Bi	84 Po	85 At	86 Rn
87 Fr	88 Ra	89-103	104 Rf	105 Db	106 Sg	107 Bh	108 Hs	109 Mt	110 Ds	111 Rg	112 Cn	113 Uut	114 Fl	115 Uup	116 Lv	117 Uus	118 Uuo

57 La	58 Ce	59 Pr	60 Nd	61 Pm	62 Sm	63 Eu	64 Gd	65 Tb	66 Dy	67 Ho	68 Er	69 Tm	70 Yb	71 Lu
89 Ac	90 Th	91 Pa	92 U	93 Np	94 Pu	95 Am	96 Cm	97 Bk	98 Cf	99 Es	100 Fm	101 Md	102 No	103 Lr

Figure 1.4 The periodic table of elements

Movement of electrons

To make the electric current, you need to move electrons from one atom to the next. To do this they need to be given a push by an external force or pressure.
The pressure used to move electrons can be created by:
• magnets
• a chemical reaction

Electrical Fundamentals

Orbiting electrons are held in place in a similar way to the gravity acting on the planets circling around the sun. Because the construction of some atoms is so simple, the attraction between the nucleus and the electrons is very strong. This makes it very hard to move electrons. When electrons don't move easily the element is known as an insulator.

A copper atom contains 29 electrons and 29 protons. The electrons orbit in circles that get bigger and bigger. The electrons in the farthest orbit have a far weaker bond/attraction to the nucleus than those in a simple atom. These outer electrons are known as 'free electrons'. If an external pressure is applied, electrons can be moved from one atom to the next. This movement of electrons is electric current. When electrons do move easily, the component made from this element is known as a conductor.

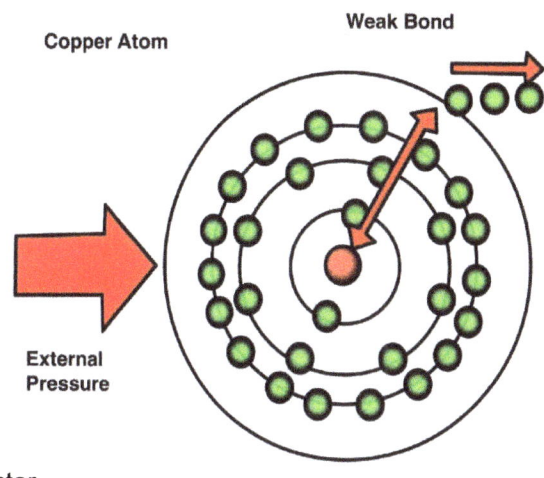

Figure 1.5 A copper atom with external pressure applied

Conductors and insulators

- Conductors are used on cars where we want electricity to flow easily, such as wiring.
- Insulators are used on cars to reduce the movement of electricity, such as the coating on the outside of a high voltage cable.

Circuits

For electrons to move from one atom to another, the conductor must be connected in an unbroken loop known as a circuit. This means that as one electron leaves it can be replaced by one from behind. If not connected in a circuit, the electrons cannot flow (move), as the last electron in the conductor has nowhere to go. If the circuit is broken it is said to have lost **continuity**.

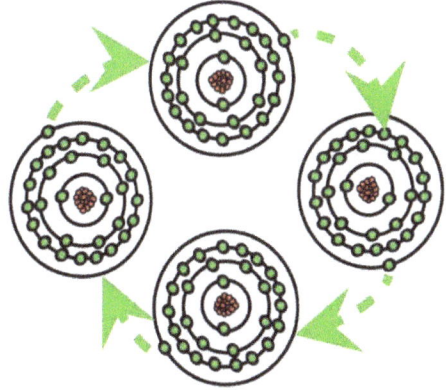

Figure 1.6 Atoms in a circle to show how electrons will move from one atom to the next

Continuity – a conductor that is complete (unbroken) which will allow the continuous flow of electricity.

Electrical Fundamentals

Magnets

Electricity and magnetism are very closely linked, and are like two sides of the same coin.
Both electricity and magnets have positive and negative, or North and South, poles. Both attract and repel.

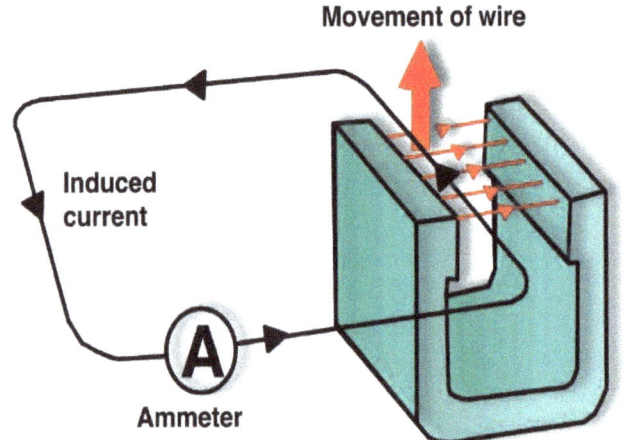

- If a copper conductor (wire) is passed by a magnet, the magnetic attraction will move electrons through that copper conductor and create electric current.
- If an electric current is passed through a copper conductor then it will generate an invisible magnetic field.

The magnetic effect of electrical current can be used to make things move (by magnetic attraction or repulsion). That movement can be used to make a motor.

The movement of magnets past a conductor can be used to make electric current; this is the principle of a generator.

- Motors turn electrical energy into mechanical energy.
- Generators turn mechanical energy into electrical energy.

Figure 1.7 The generation of electrical energy by moving a wire through a magnetic field

Chemical reaction

Electrical energy can be produced by or converted into chemical energy. Because of this it is possible to store electricity and take it with you in the form of a battery. If you keep a battery charged, it provides a portable source of electricity that can be used when needed.

The principle of a direct current circuit

As it is very hard to imagine electrons moving from one atom to another, the process is often best described using water as an analogy.

Imagine a simple water tower containing:
- a reservoir of water at the top (to represent a battery)
- a pipe leading from the bottom of the reservoir (to represent the wire)
- a tap on the end of the pipe (to represent a switch)
- a small water wheel at the end (to represent a motor)

This analogy can be used to show the operation of a simple electrical circuit. When the tap is turned on, gravity pushes water down through the pipe, under pressure, out through the tap and on to turn the water wheel. This is like the way electric current flows through a circuit and turns a motor when the switch is turned on.

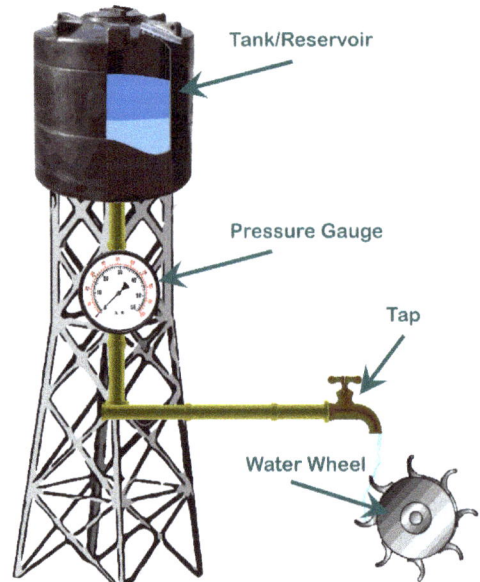

Figure 1.8 A water tower used to represent a direct current (DC) electric circuit

Electrical Fundamentals

- The quantity of water flowing through the pipes, is like the amperage or current flowing in an electric circuit.
- The pressure of water within the pipes is like the voltage in an electric circuit.
- The slowing down of water flow caused by the tap is like resistance in an electric circuit.
- The rate at which the water wheel works is like power in an electric circuit.

Series and parallel circuits

Two main types of electrical circuit are used in the construction of motor vehicles:

- Series
- Parallel

Series circuit

In a **series circuit** the consumers are connected in a line one after another. Because they are all in the same circuit, they share the electricity provided depending on the amount of power that they use. If more than one **consumer** is fitted it will only get part of the voltage available. (The more powerful the consumer, the greedier for electricity it is).

Series circuit – a circuit with electrical consumers connected in a line, one after another.

Consumer – an item or component that uses up electrical energy, i.e. bulbs, motors etc.

Parallel circuit – a circuit where electrical consumers are connected side by side.

If any one of the consumers fails, the circuit is broken and no electricity can flow. The rest of the consumers stop working. This makes series circuits unsuitable for many systems on cars.

Parallel circuit

In a **parallel circuit** the consumers are connected next to each other. Each has its own power supply and earth return, back to the battery.

Because each consumer has its own supply and earth, all the consumers receive the full voltage available, and work at full power.
If one consumer in the circuit fails, the others keep working.

Ohm's law

If any one of the units within a circuit (volts, amps, ohms or watts) is changed (i.e. increased or decreased), this will affect all the other units. Using a water analogy:

- If the voltage (or pressure) in a water system was increased, more water would flow and the amperage (or quantity) would also increase.
- If the resistance to flow was increased (if a tap was partially closed, for example) then less water would flow and the amperage (or quantity) would fall.

Electrical Fundamentals

This was explained by Georg Ohm with the following mathematical calculations:

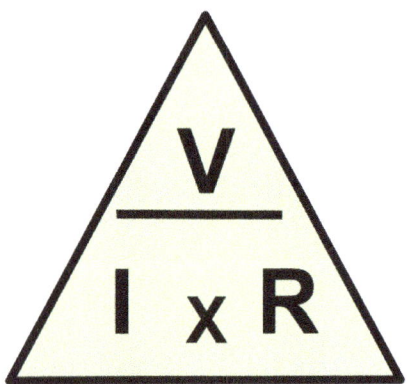

amps = volts ÷ resistance
resistance = volts ÷ amps
volts = amps × resistance

With Ohm's law, if you know two of the electrical measurements, you can calculate the third.
The Ohm's law triangle is a good method for calculating the missing unit. It is laid out as shown in Figure 1.9.

Figure 1.9 Ohms law triangle

In Figure 1.9:

- V = volts (this is sometimes shown as the letter 'E' to represent EMF, but still means volts).
- I = amps (the letter 'I' is used to represent instantaneous current flow).
- R = ohms (the letter 'R' is used for resistance because an 'O' could be confused for a zero).

How to use the triangle
Cover up the unknown unit with your thumb and you are left with the calculation required. For example, amperage is unknown, so cover the 'I' and you are left with V ÷ R (i.e. volts divided by resistance).

Using Ohm's law to help diagnose faults

The relationship between voltage, resistance and amperage can help you to diagnose faults within an electrical circuit. If you take measurements using the different electrical units and then compare them using the Ohm's law calculation, you will be able to work out if the fault is occurring because of:

- **Pressure (volts)**
- If this is lower than expected, component performance is reduced.
- If this is higher than expected, component damage can occur.

- **Quantity (amps)**
- If this is lower than expected, component operation will normally be incorrect.
- If this is higher than expected, component/system operation is being overworked.

- **Resistance (ohms)**
- If this is lower than expected, current may be taking an alternative path to earth (short circuit).
- If this is higher than expected, it will consume electrical energy and reduce system performance.

Electrical Fundamentals

The power triangle

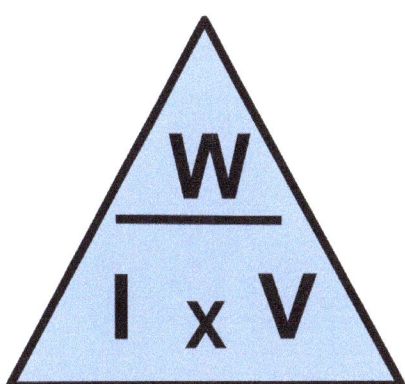

Watts or power can be calculated in a similar way as:

amps = watts ÷ volts
volts = watts ÷ amps
watts = amps × volts

A power triangle can be used in the same way as Ohm's law. It is laid out as shown in Figure 1.10.

Figure 1.10 Power law triangle

In Figure 1.10:

- W = power (in watts – this is sometimes shown as the letter 'P' to represent power, but still means watts).
- V = volts (this is sometimes shown as the letter 'E' to represent EMF, but still means volts).
- I = amps (the letter 'I' is used to represent instantaneous current flow).

How to use the triangle
Cover up the unknown unit with your thumb, and you are left with the calculation required. For example, amperage is unknown, so cover the 'I' and you are left with W ÷ V (i.e. watts divided by volts).

Common electrical faults

Diagnosing electrical faults can be confusing as the symptoms can be very wide and varied. If you follow a simple approach, the diagnosis can be reduced to four main electrical faults:

- Open circuit
- High resistance (including bad earth)
- Short circuit
- Parasitic drain

Open circuit

In an open circuit, electricity cannot flow. This is normally because there is a physical break in the system. As a potential difference (Pd) will only occur in a circuit when current can flow, the fault can be diagnosed using a test method known as volts drop.

> To diagnose an open circuit fault, voltage will be seen before the break, but no voltage will be measured after the break.

Electrical Fundamentals

High resistance

In a high resistance circuit, the electricity slows down. This is normally because of a partial restriction in the system. Many high resistance faults are caused by poor, corroded or loose connections. A potential difference (Pd) will occur in a high resistance circuit, but the total voltage will be shared between the consumer and high resistance. This fault can also be diagnosed using the test method known as volts drop.

> To diagnose a high resistance fault, full voltage will be seen before the resistance, but a lower voltage will be measured after the resistance.
>
> A high resistance will make circuit current fall and this can sometimes be seen if an inductive amp clamp is connected to the circuit, and the amplitude value on the oscilloscope screen is compared with the fuse rating.

Bad earth

A bad earth is a high resistance after the consumer. If this exists, the symptoms will be that the component won't work properly. Sometimes a bad earth can also cause the electrical energy to find an alternative path to the negative side of the battery through another circuit, causing unexpected operation of other consumers.
To diagnose a bad earth, use the same procedure as for high resistance.

Short circuit

Electricity is lazy, and will always take the path of least resistance. (Why travel the full length of the circuit when it can take a shortcut?)
In a short circuit, the electricity doesn't make it all the way to the end. Instead of going through the consumer, the electricity makes its way back to the battery early, and in the process, converts its energy to heat.
The sudden discharge of current can cause a lot of damage, so the fuse that is used to protect the system should blow. If this happens the symptoms can make you think that the problem is an open circuit (which it is in a way, as the blown fuse has broken the circuit so that no current can flow).
In this situation, you can test the system in the same way as explained in testing an open circuit, but once you have discovered the blown fuse, you should change your diagnostic routine to look for a short circuit. Any heat damage, including blown fuses, is a good indication that a short circuit might exist.

If a **dead short** to earth exists (e.g. the insulation of a wire has chafed against the metal bodywork of the vehicle for example), you can use a test lamp to help diagnose this fault. It is important to use a test lamp containing a bulb and not an LED, as this could lead to system damage.
Once connected in place of the fuse, if the test lamp illuminates then the electricity is finding an alternative path back to the battery (short circuit). As the bulb is an electrical consumer, it uses up the electrical potential, and shouldn't damage the rest of the circuit. The circuit should then be disconnected systematically from the far end, working back towards the fuse box. When the bulb goes out, you have located the position of the dead short.

> **Dead short** – an electrical short circuit that goes straight to earth without passing through a consumer.

Electrical Fundamentals

A blown fuse may also be an indication of excessive current draw. With a test lamp connected as described in the short circuit test, an inductive amp clamp can be attached to the circuit (around the test lamp wire) and real time measurements can be taken by switching components on and off until the one with the high current draw is found.

Parasitic drain

A parasitic drain is similar to a short circuit – electricity will continue to flow even if the system is switched off, although this fault may not cause visible system damage. The symptom reported is normally that the battery goes flat if left for a period of time. To help diagnose this fault, you can use the inductive clamp meter as an ammeter. (This acts like a flow gauge to measure the amount of electric current moving in a circuit).

Checking a parasitic drain

To check for parasitic drain, switch off all electric systems and connect the current clamp.
With everything switched off, there should be no (or very little) current on the display of the scope. If any current (measured in amps) is shown, then a parasitic drain exists. To help find the parasitic drain, remove the fuses one at a time until **amps draw** falls to zero. This will help you locate the circuit containing the drain. Once you have identified the circuit, disconnect the components in that system until the current draw falls once again. You can now replace the faulty component.

Amps draw – the amount of current being used.

To help identify the amount of current a fuse is capable of withstanding before it blows, they are often colour coded. An example of the standardised colour codes is shown below:

Tan 5 amps

Brown 7.5 amps

Red 10 amps

Blue 15 amps

Yellow 20 amps

Clear 25 amps

Green 30 amps

Electrical Fundamentals

DIN terminal numbers

To help technicians identify electrical terminals on vehicle circuits they are often given a standardised number. (For examples of these terminal numbers, see appendix in the back of this book).

Breakout boxes

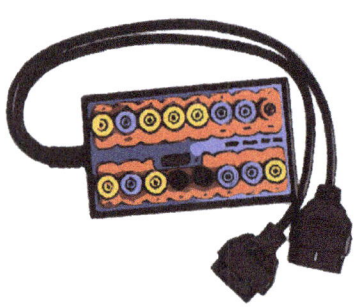

To help with electrical diagnosis, some manufacturers produce breakout boxes. A breakout box or wire is a piece of test equipment that enables the technician to measure electrical values at terminals or connectors without disconnecting the circuit. The breakout box fits in series with an electrical connector in the circuit to be tested and has a cable connected in parallel leading to a 'pin-out'. The uninterrupted circuit can now be operated as normal and live measurements taken at the pin-outs using an oscilloscope.

Figure 1.11 A breakout box

Electronic terminology

Table 1.2 describes some electronic terminology used in automotive systems.

Table 1.2 The operation of electrical and electronic systems

Electrical/Electronic system component	Purpose
ECU	The electronic control unit is designed to monitor the operation of vehicle systems. It processes the information received and operates actuators that control system functions. An ECU may also be known as an ECM (electronic control module).
Sensors	The sensors are mounted on various system components and they monitor the operation against set parameters. As the vehicle is driven, dynamic operation creates signals in the form of resistance changes or voltage which are sent to the ECU for processing.
Actuators	When actioned by the ECU, motors, solenoids, valves, etc. help to control the function of the vehicle for correct operation.

Electrical Fundamentals

Table 1.2 The operation of electrical and electronic systems

Digital waveforms

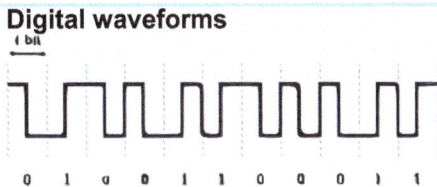

Many vehicle sensors create analogue signals (a rising or falling voltage). The ECU is a computer and needs to have these signals converted into a digital format (on and off) before they can be processed. This can be done using a component called a pulse shaper or Schmitt trigger.

Analogue waveforms

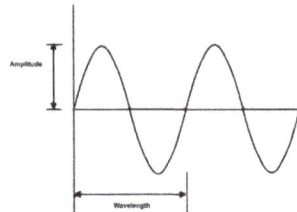

An analogue waveform is often created by inductive sensors. It is normally shown as a rising or falling voltage on an oscilloscope screen. This alternating current (AC) is also known as a sinewave.

Duty cycle and PWM

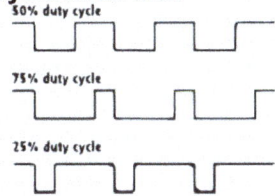

Lots of electrical equipment and electronic actuators can be controlled by duty cycle or pulse width modulation (PWM). These work by switching components on and off very quickly so that they only receive part of the current/voltage available. Depending on the reaction time of the component being switched and how long power is supplied, variable control is achieved. This is more efficient than using resistors to control the current/voltage in a circuit. Resistors waste electrical energy as heat, whereas duty cycle and PWM operate with almost no loss of power.

Networking and multiplex systems

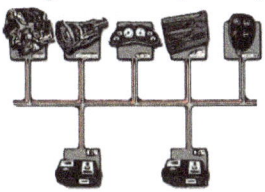

Many modern vehicle systems are controlled using computer networking. In these systems a number of ECU's are linked together and communicate to share information in a standardised format.

Chapter 2 Introduction to Oscilloscopes

This chapter will help you develop an understanding of the set-up and operation of automotive oscilloscopes. It describes how they can be used for electrical and electronic diagnosis, and the technical vocabulary that supports their functions. The content is designed to get you up-and-running as soon as possible, with straightforward descriptions, allowing you to develop your knowledge and understanding in a practical environment. Remember to work safely at all times and observe the relevant environmental, health and safety regulations, while developing diagnostic routines that are systematic and effective.

Contents

Good practice and systematic diagnostic routines 21

What is an oscilloscope? 21

Oscilloscope set-up and use 22

Waveforms and terminology 27

There are many hazards associated with the service and maintenance of light vehicle electrical and electronic systems. You should always assess the risks involved with any diagnostic, maintenance or repair routine before you begin and put safety measures in place.
You need to give special consideration to the possibility of:
• The risk of electric shock.
• The hazards associated with running engines in confined spaces.
You should always use appropriate personal protective equipment (PPE) when you work on these systems. Make sure that your selection of PPE will help protect you from these hazards.

Don't forget your PPE and VPE

  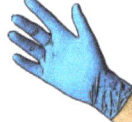 

Introduction to Oscilloscopes

Good practice and systematic diagnostic routines

Routes to diagnosis

You need to develop a logical and systematic approach to your diagnostic routine in order to cut down on time, cost and frustration. Remember that correct diagnosis is important to ensure a first-time fix.

The flow chart shown in Figure 2.1 gives an example of a generic routine that should be used when diagnosing a fault.

Figure 2.1 Generic diagnostic routine

What is an oscilloscope?

An oscilloscope is a piece of electrical test equipment designed to show measurement over a period of time. Instead of taking a static reading and giving a straight numerical value for the result, an oscilloscope draws a picture or creates a signature of measured results and displays this as a graphical representation.

For many years, oscilloscopes have been used in electrical engineering and medicine, however, the automotive industry also benefits from its outstanding diagnostic capabilities.

Hospitals use medical oscilloscopes to monitor heart function, and by interpreting the images produced on a screen or print-out, can make sound diagnostic decisions about the state of someone's health; it does this by measuring electrical impulses produced through contacts placed on the patients' body. An automotive oscilloscope functions in the same manner and can be thought of as a vehicle heart monitor, with electrical contacts connected to various components on the car.

The advantage of using oscilloscopes to monitor vehicle function is that live measurements can be taken while the system is in operation and this extends the scope of diagnosis to test components under various operating conditions.

If you are looking to purchase your first oscilloscope, it can be a good idea to buy a simple, cheap and uncomplicated scope initially and get used to using it for straightforward diagnosis on a regular basis. This way, some of the issues that can arise from the perception that oscilloscopes are complicated and unnecessary can be reduced.

A cheap scope is more likely to be left out and ready for use, as its general value means that it might not be treated with the same sort of reverence as one that has cost a great deal of money and investment.

If a scope is used on a regular basis, familiarity will ensure that becomes a key tool in the diagnosis of complex vehicle faults, resulting in confidence in its use and abilities.

Once your ability has overtaken the capabilities of the tool, it may then be appropriate to invest in a more expensive piece of equipment that now suits your needs.

Introduction to Oscilloscopes

Types

Oscilloscopes come in all shapes and sizes, with functions ranging from simple to complex. As a general rule, 'you get what you pay for'.
A cheap oscilloscope will often be quick and easy to set-up, but may not have the functions and capabilities of a more expensive piece of equipment.

It is not uncommon for some organisations to own several different types of oscilloscope that can be used in slightly different ways, depending on the complexity of the diagnostic repairs that present themselves. For example, a small cheap scope is used during the initial diagnosis, keeping the operation quick and uncomplicated and then if required, the more expensive and capable machine can be utilised if the complexity of the task requires it.

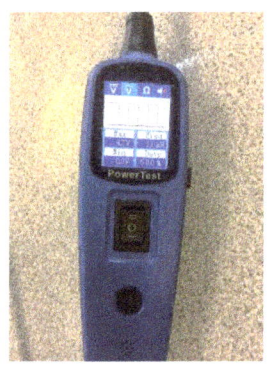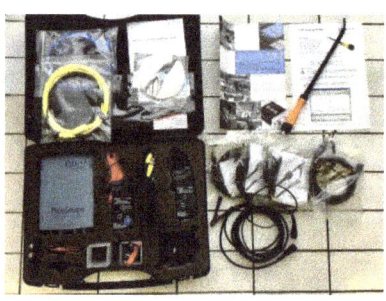

Figure 2.2 Different styles of oscilloscope

Set-up (standard voltage measurement)

The screen of an oscilloscope will display a graph which represents measurements taken over a period of time; this gives the opportunity to view the function of components or systems during normal vehicle operation.
As with all graphs, the display will show a minimum of two axis.

- The vertical axis (Y) describes the amplitude of the measurement – the amount of volts, current (amps) or pressure (if used with a pressure transducer adapter).
- The horizontal axis (X) describes the timescale measurement – the period of time (sometimes known as the sweep) over which the readings are taken.

Once powered-up, the oscilloscope should display a line moving horizontally across the graph from left to right, representing the live value being measured and leaving a trail behind to show a history of the measurements received.

Introduction to Oscilloscopes

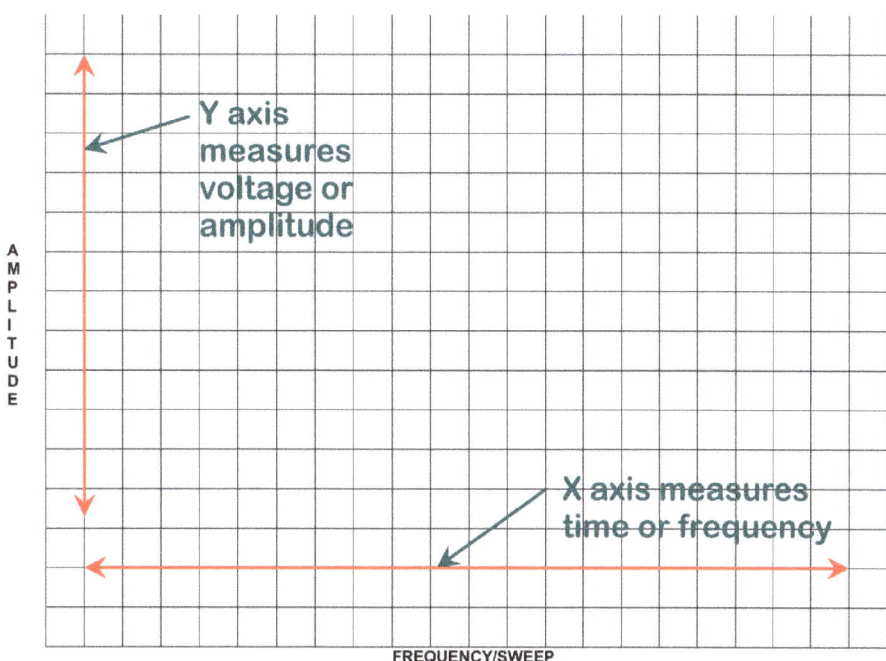

Figure 2.3 An oscilloscope screen

Depending on the design of your oscilloscope, a method of adjusting the values for amplitude and sweep will be available, either as dials, buttons or drop-down menus. At this point it is not necessary to make adjustments to these settings.

The oscilloscope will be supplied with an appropriate set of test leads, and theses should now be connected to the machine at the correct sockets. It is often the case that sockets are colour coded to make connection straightforward, however, it may require reference to the manufactures operating manual.

Two types of connector are common:
- Socket type (sometimes known as banana connectors)
- BNC – a quick twist fit connector named after its inventors (Bayonet Neill–Concelman)

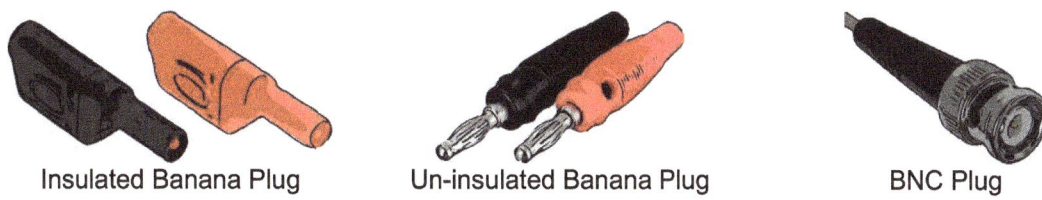

Figure 2.4 Test lead connectors

Introduction to Oscilloscopes

No matter which type of connector is supplied/used with your oscilloscope, adapters are often available which can be used to attach various accessories from other manufactures.

Once attached to the scope, the test probes can now be connected at the vehicle component to be tested.
The probes will often have an earth or ground lead and this should be connected to a good source of ground or earth on the vehicle (the battery negative post is the best place, and is often easily available/accessible).
The remaining test probe can now be used to connect at the circuit of the component being examined.

With the vehicle system being operated, adjust the amplitude and timescales until a waveform appears on the screen.

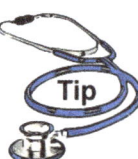

If you don't know the scale for voltage or sweep to use, simply select a mid-range value available on the oscilloscope, and move up or down as required, until the height and width of the pattern can be easily viewed in full.

The measurement probes of a scope should be passive, only receiving voltage inputs from the circuit being tested; as they don't put out any voltage, it is safe to experiment with scale settings, as long as high voltage circuits are not being tested.

If you are testing high voltage or ignition circuits, you must use the correct circuit adapters and take appropriate precautions.

If you are not aware of the hazards involved when testing these systems, you increase the risk of personal injury or damage to vehicle systems and equipment.

Uses

An oscilloscope can be used to test vehicle electrical systems in detail, but should never be confused with the graphing meter function found on many scan tools.
A graphing meter is a method of converting live serial data into a graphical representation of sensor and actuator parameters (PID's).
The main difference here is that an oscilloscope will take immediate true readings from components in real-time, whereas, a graphing meter is showing an interpretation via the ECU, of data that has been gathered and processed (i.e. second-hand information).

An oscilloscope can be used to measure the pulse of a vehicle component by showing an image of voltage or amperage (or both if used with a multi-channel setting), which is very useful when checking for the correct function or 'health' of a system.
A single channel will draw the signature of the specific component being tested, while multiple channels will show the operating relationship between components. It is capable of measuring all switched or variable electrical components on a vehicle, however, it is not recommended for direct use on SRS, or high voltage drive systems unless you have had specific training and knowledge.

Introduction to Oscilloscopes

To take measurements, the oscilloscope will often be provided with a series of accessories which help you access the various system components. The more comprehensive the scope, the more accessories are often available, and this will mean that a large case full of wires and connectors is provided. Unfortunately, this promotes the misconception that the equipment will be difficult to set-up and use.

Many accessories will be test leads, probes and clips to help you connect the scope to the desired component/circuit, however, some have specialist functions and are designed specifically for purpose. Examples of diagnostic accessories can be seen in table 2.1.

Table 2.1 Oscilloscope diagnostic accessories

Accessory	Description
Probes and back probes 	Many oscilloscopes have a selection of interchangeable probes with different styles of contact point. Some probes may have a very thin point that can be used to '**back-probe**' a connection by pushing it into the rear of the components' electrical connectors so it can touch the terminal and test the signal without causing damage to the wiring or circuit.
Crocodile clips 	Crocodile or alligator clips are small clamps, often manufactured with teeth on the inner jaws, which can be fitted to the end of the test leads and connected to the circuit or component to provide hands-free operation.
Secondary ignition connectors 	In order to test the high voltage produced by a secondary ignition system, adapters are available which clip around the high-tension lead and detect the pulse created when a spark is produced, converting it to a waveform for display. As many modern ignition systems no longer have high-tension leads, it may not be possible to use these connectors without an additional extension lead adapter which fits between the ignition coil and the spark plug.
Inductive secondary ignition probes 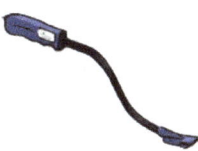	If an ignition system is designed as coil on plug (COP), some scope manufacturers provide secondary ignition probes that are **inductive**. These probes vary in shape and design, but can simply be held against the high-tension component being tested and produce a waveform for display.
Fused circuit loop adapters 	This is a small loop of wire which is connected in series at the fuse box and can then be used to create a location where an inductive amp clamp can take readings from that circuit. Although the fuse has been removed, there is an inline fuse holder in the loop itself, and this is used to help protect the circuit while testing is conducted.

Introduction to Oscilloscopes

Table 2.1 Oscilloscope diagnostic accessories

Accessory	Description
Inductive amps measurement clamps 	An inductive amp clamp is a vital accessory if you want to measure current using your oscilloscope. Unlike a standard ammeter, which must be connected in series with a circuit, an inductive clamp simply clips around a wire of the circuit that you want to test, the circuit is switched on and the readings taken. The inductive clamp will have its own internal power source (battery) and in many cases, uses the Hall effect principle to sense the magnetic field created around a wire when the circuit is switched on. The strength of the magnetic field is then converted into a waveform displayed on the oscilloscope screen and is interpreted as amperage.
Attenuators 	An **attenuator** is a resistor component which is connected in series with the test lead and probe. It is designed to allow the oscilloscope to measure higher voltage values than those calibrated on the readout screen. They will normally be designed to reduce the displayed amplitude in multiples of ten (i.e. x10, x20 etc.) and this will be marked on the casing of the attenuator. If an attenuator is used during your testing routine, you must multiply the voltage readings displayed on the graph by the value shown on the attenuator.
Pressure transducers 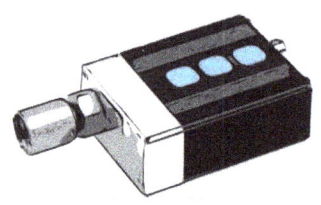	A pressure **transducer** is an accessory which, with a set of adapters, can be connected to vehicle systems or components and record physical measurements of pressure for either fluid, gas or vacuum. When pressure is sensed, it is converted into an electrical signal that can be interpreted as an amplitude on the display of the oscilloscope.

Back-probe – an electrical connection made at the component plug which contacts the terminals without damage to the wirings insulation.

Inductive – a process where a voltage is created inside a conductor by magnetism.

Attenuator – a component accessory used to reduce the strength of a signal recorded for testing.

Transducer – a device that converts one form of energy to another; pressure into an electrical signal for example.

Introduction to Oscilloscopes

If you cannot access the electrical circuit using a breakout, the preferred method of connection for the oscilloscope is by 'back-probing' the terminals at a connector plug.
Piercing the insulation of a wire will expose the copper core to the effects of oxidation, causing corrosion and resistance if left open. If wire piercing probes are used, make sure that the wire is effectively re-insulated.

Waveforms

The problem found testing modern vehicle electronic systems using a multimeter is that they are not quick enough to keep up with any rapid changes of voltage or current; the display cannot refresh fast enough.
An oscilloscope draws an image of voltage or current presented over a period of time, giving a picture of what is happening inside an electric circuit. It is like having a route map, showing the direction the electricity has travelled. This image, drawn on the display of an oscilloscope, is known as a waveform.
Waveforms create the signature of the component being tested, and with some practice, a large amount of diagnostic information can be gained from analysing these signatures.

Waveforms are displayed on a graph with an 'X' and 'Y' axis.
Amplitude
Amplitude is measured vertically up the side of the graph on the 'Y' axis. Depending on the type of measurement being taken, it displays the amount of:
Voltage
Amperage/Current
Or pressure (if using used in conjunction with a pressure transducer)
Sweep
Sweep is measured on the 'X' axis, horizontally across the bottom of the graph. This is the time measurement, which may also be labelled frequency

Amplitude – the height of the waveform.

Sweep – the width of the waveform.

.
Current, voltage and pressure measurements

Voltage

To measure voltage using an oscilloscope, simply connect the test probe leads supplied to the correct sockets (as per manufacturer instructions), attach the ground lead to a good source of earth and the test probe to the component or circuit that you want to test. Some oscilloscopes might need to be set to a voltage test probe in the menu section (and it may also be necessary to say how many channels you will be using).
When the circuit is operated, if a pulse is seen that affects the line of the waveform, the amplitude and sweep should now be adjusted so that the pattern is clearly displayed on the screen. (Amplitude will adjust the height of the waveform, and sweep will adjust the width).

Introduction to Oscilloscopes

Amp clamp

To measure current using an oscilloscope, you should attach an inductive amp clamp. Inductive clamps are often not supplied with the original oscilloscope kit and may have to be purchased as a separate accessory. The inductive amp clamp is normally connected to the scope via the same ports as those used for voltage probe measurement. This is because an inductive amp clamp is not measuring true amperage, but is instead using a method to pick up the electromagnetic interference given off by current flowing through a wire, and the voltage sensed from this is converted to an amplitude on the screen which can be interpreted as current flow. Now the clamp has been connected, to the scope, you simply have to clip it around a wire of the circuit to be tested.

The amp clamp will have a plus and minus symbol on the clamp body to show which way round it should be connected to a circuit, however, it is not necessarily important to know this; if the clamp is connected with the wrong polarity, it will simply invert the image and the waveform will appear upside-down.
If this happens, you can turn the clamp round the other way.

An inductive clamp needs its own internal power source, normally provided by a 9 volt battery or similar, and will need to be switched on before measurements are taken. The on switch will often have a choice of sensitivity, and depending on the amount of current you will be testing, you should select the correct scale. If you have chosen the incorrect scale, once you begin testing, simply alter the switch and continue your test.
Due to the nature of the inductive amp clamp picking up on electromagnetic interference, it will need to be calibrated before use. This can be achieved by zeroing the clamp using a dial or push button provided by the manufacturer. Once the clamp has been calibrated, the circuit can now be switched on and measurements taken. When the circuit is operated, if a pulse is seen that affects the line of the waveform, the amplitude and sweep should now be adjusted so that the pattern is clearly displayed on the screen. (Amplitude will adjust the height of the waveform, and sweep will adjust the width).

Pressure transducers

Pressure transducers are an accessory that can be used to convert a physical pressure measurement into an electrical impulse that can be displayed on an oscilloscope screen. These accessories are rarely supplied with an oscilloscope and must be purchased separately.

Pressure transducers come in different sizes and shapes depending on what you want to test, and therefore it may be necessary to purchase a variety in order to carry out all of the functional testing that you may wish to do.

The pressure transducer should be connected to the oscilloscope following manufacturer's instructions, and then attached to the system from which the pressure is to be measured. The system can then be operated or pressure simulated so that a waveform can be seen on the screen of the scope.
When the circuit is operated, if a pulse is seen that affects the line of the waveform, the amplitude and sweep should now be adjusted so that the pattern is clearly displayed on the screen. (Amplitude will adjust the height of the waveform, and sweep will adjust the width).

Introduction to Oscilloscopes

If you wish to hold one repeating pulse of any waveform steady on the display, you need add a trigger.

Triggers

When the trace of a waveform reaches the right-hand side of the display, it stops and begins again back on the left-hand side of the screen; this is known as 'triggering'.

A trigger is a mechanism provided by oscilloscope manufacturers which allows you to keep the image from a waveform pulse steady on the screen by setting the point at which the screen starts again (i.e. its 'triggering point'). With the old-fashioned cathode ray tube (CRT) lab scope, it was necessary to attach extra wires or adapters for the trigger, but with modern digital scopes this option is programmed and calculated digitally.
Once the trigger is activated from the menu, a marker such as a small cross, diamond or square will appear somewhere on the scope display. This marker must then be moved to a chosen position on the screen by arrow buttons or in the case of touch screen display 'drag and drop'.
As long as the trigger marker is somewhere within the waveform pattern itself for reference purposes, it will display the pulse shown at a fixed position making it easier to see and interpret.
Another function that is often available with triggers is the ability to choose whether the pulse is held on a rising or falling slope of the waveform. This gives you the ability to place the waveform on the screen in the position that gives you the clearest view of the section you want to see.

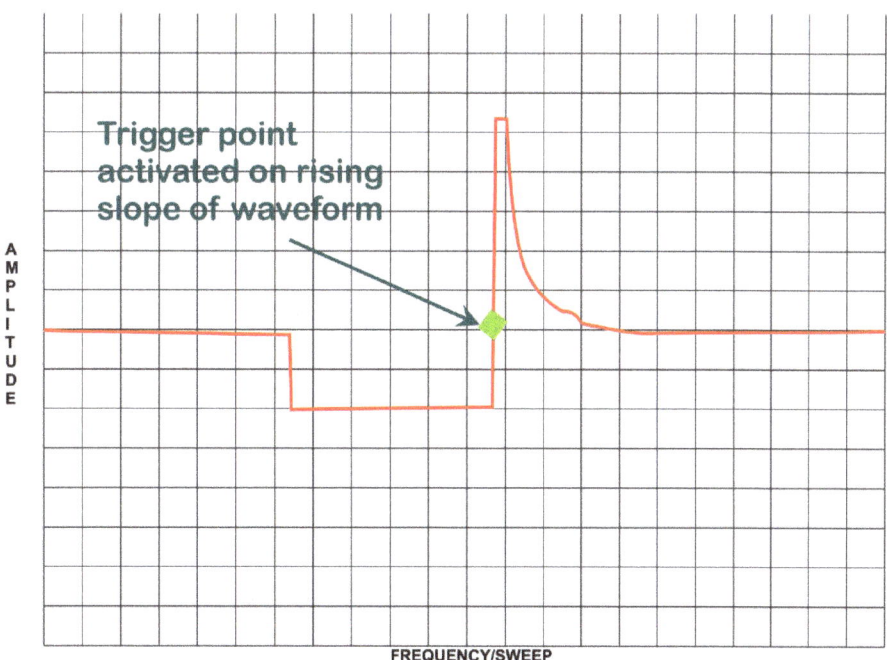

Figure 2.5 Trigger point

Introduction to Oscilloscopes

Channels

Some automotive oscilloscopes come with multiple channels, which is the name given to the amount of separate input connections on the scope, and how many waveforms it can display at once. Channels give the user the opportunity to test and display more than one circuit or component at the same time.
A single channel gives the operator the ability to analyse the signature produced by a particular component or circuit, whereas multiple channels allow the user to compare components and how they interact.

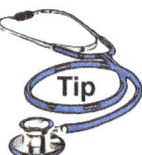

If your oscilloscope has the capability of flight recording, you can use multiple channels to help isolate the primary cause of an intermittent failure.
By setting up the scope and connecting it to a selection of suspected components, the system can be operated until it cuts out or stops working correctly.
The display can then be paused, and by scrolling back through the screens, you will be able to see which component was the first to stop working; causing the others to follow (a 'domino effect').

Filters

Oscilloscopes are so sensitive that they often give a very noisy waveform because they pick up on electromagnetic interference from the surrounding environment. Some scope manufacturers provide the option to apply a filter to the measured waveform which artificially smooths out the pattern, making it easier to read.

Remember that if the filter option is used, the quality of the waveform pattern is reduced and this could disrupt the amount of diagnostic information available.

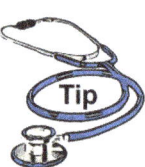

The addition of a filter can be useful when a trigger is finding it difficult to precisely hold an image steady on the display, as noise in the waveform disrupts the triggering point.
Multiple rising or falling edges to the waveform can lead to the trigger creating a situation known as 'ghosting', where the pattern becomes fuzzy, due to a number of overlapping waves appearing at the same time.

Cursors

Another function that's available on some oscilloscopes is the ability to add cursors to the graph and take measurements at certain points. Once the cursors are activated, they project lines either horizontally or vertically across the graph which can be positioned to highlight certain sections of the waveform. When aligned, a textual display will show the exact position of the cursor lines and helps provide information about amplitude, and time.

Introduction to Oscilloscopes

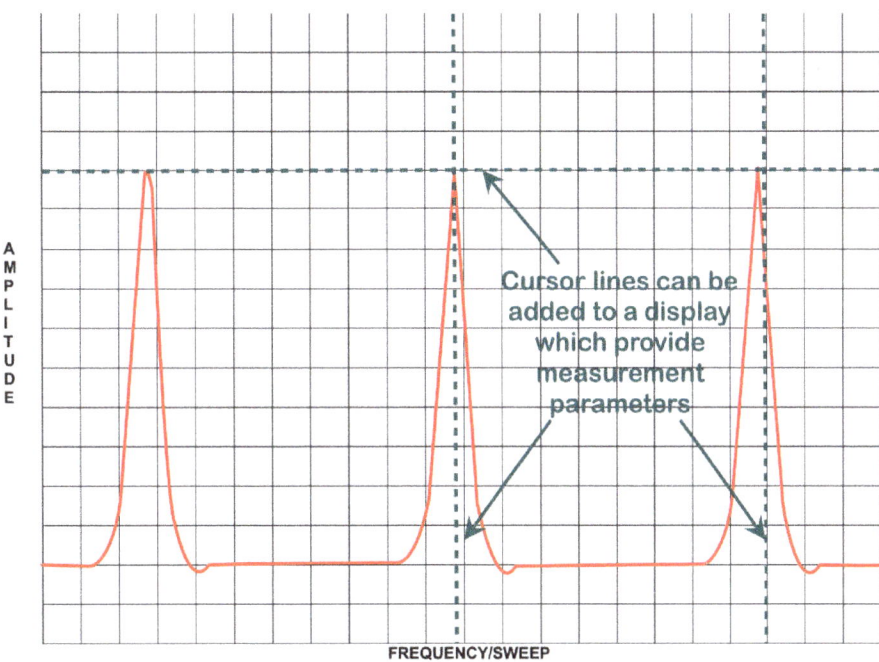

Figure 2.6 Cursor lines

Delta and frequency measurements

If two cursors are used vertically on the screen of an oscilloscope, once their position is set, the display can indicate two different time measurements known as delta and frequency.
- Delta is the time measurement between the two cursor points.
- Frequency is how many complete cycles have occurred in one second and is measured in Hertz (Hz).

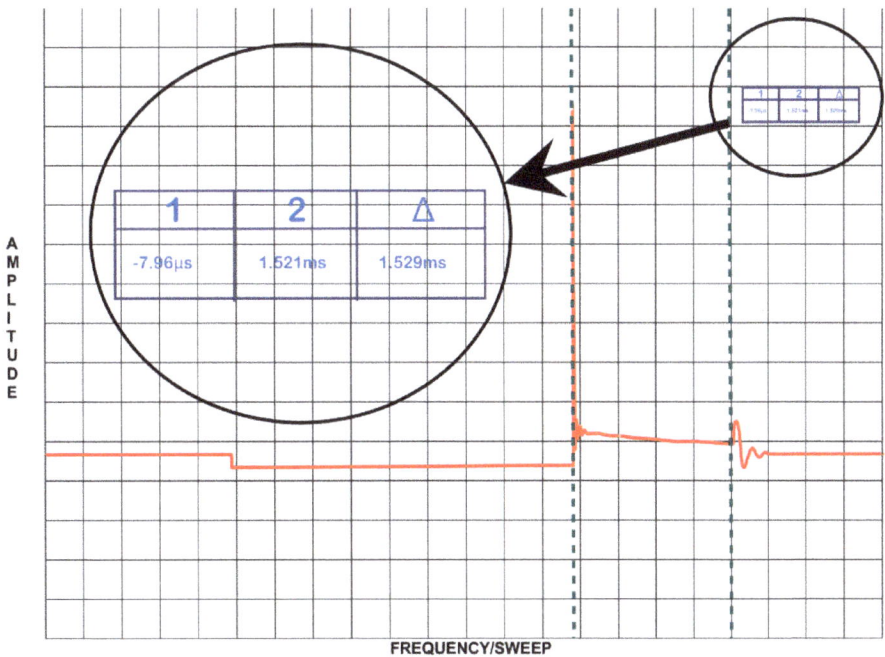

Figure 2.6 Delta and frequency image

Introduction to Oscilloscopes

Table 2.2 shows a helpful summary of oscilloscope functions (including pre-sets and pause/play, sometimes offered by oscilloscope manufacturers).

Table 2.2 Oscilloscope functions

Function	Purpose
Cursors	A function on some digital oscilloscopes is the ability to add cursors to the waveform screen and take measurements. Once the cursor option has been selected, depending on whether the measurement is to be taken on an 'X' or 'Y' axis, the cursor can be moved so that they match a point on a paused waveform. Cursors can be used to measure the time or the amplitude, and the readings are often displayed in a box or chart on the screen of the oscilloscope. If two cursors are used on the same axis, the display will normally register both positions and the difference measurement between them; this is known as the 'delta'.
Triggers	Often, if a waveform shows a repeating pattern, it will move rapidly across the screen at the speed of the sweep (time base). This can make it difficult to analyse a particular pattern unless the screen is paused. Once the screen has been paused, you are no longer observing live readings. Many scopes have the ability to add a trigger to the display which will anchor the pattern so that it repeats at the same position over and over again. This gives the impression that the pattern has been paused, but it is actually a live reading that continues to be displayed. When activated, the trigger displays a small indicator on the screen, which must be positioned somewhere within the pattern of the waveform, and the image will now be held steady. On some oscilloscopes, there will be the option to set whether the trigger operates on either a rising or falling edge of the displayed waveform, giving you the choice of how it will be positioned.
Pre-sets	Some automotive oscilloscopes come loaded with pre-sets. These are settings that have been added to the memory of the equipment and allow fast set-up by the operator. Normally chosen from a menu, pre-sets give options relating to vehicle components and when selected, automatically pick the time base and amplitude for given tests. Some oscilloscopes offer the option of saving your own settings as pre-sets which can be recalled at a later time. Due to the variety of signals produced by different manufacturers, pre-sets should never be relied on completely to give an accurate set-up first time; the amplitude and frequency may need to be adjusted so a clear waveform is displayed.
Pause/play	The ability to review and check a recorded set of readings on a digital oscilloscope enhances the diagnostic capabilities of the tool. Digital scopes are often able to 'flight record' a measurement session and then the operator can pause the recording and play it back at their leisure. Once paused, there will be the option to rewind or move forward through the recording, either as continuous play or as separate frames. At any point the oscilloscope display can be restarted and the screen returned to live readings.
Filters	Due to the sensitivity and accuracy of most oscilloscopes, it is not unusual for interference to be displayed as part of a measured waveform. This can sometimes make interpretation of the patterns difficult. A filter option is provided by the manufacturers of some oscilloscopes which when activated will smooth out the waveform being observed and give a clearer view. Care must be taken when using this option and the operator should remember that this is now an artificially enhanced pattern with some of the actual detail lost.

Automotive Actuators & Waveform Analysis

Chapter 3 Automotive Actuators and Waveform Analysis

This chapter will help you develop knowledge and understanding of automotive actuators. It will enable you to conduct effective diagnosis and repairs of system faults; supporting you by providing a breakdown of waveform images and analysing why the patterns are formed. Remember to work in a systematic way, and observe the relevant environmental, health and safety regulations at all times.

Contents

Actuators and their purpose	35
Solenoids and rotating motors	40
EVAP purge valve	41
Glow plugs	43
EGR valve	45
Fuel pump	46
IAC valve	48
Petrol injectors	51
Diesel injectors	58
Quantity control valve	65
Throttle servo motor	66
Cooling fan	68
VVT control	69
Starter motor	71
Alternator	75
ABS solenoid	81

There are many hazards associated with the service and maintenance of light vehicle electrical and electronic systems. You should always assess the risks involved with any diagnostic, maintenance or repair routine before you begin and put safety measures in place.
You need to give special consideration to the possibility of:
• The risk of electric shock.
• The hazards associated with running engines in confined spaces.
You should always use appropriate personal protective equipment (PPE) when you work on these systems. Make sure that your selection of PPE will help protect you from these hazards.

Automotive Actuators & Waveform Analysis

Don't forget your PPE and VPE

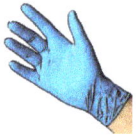

Information sources

The complex nature of light vehicle electric and electronic systems requires a good source of technical information and data. In order to conduct diagnostic, maintenance and repair procedures, you need to gather as much information as possible before you start.

Sources of information may include:

Table 3.1 Possible information sources

Verbal information from the driver	Vehicle identification numbers
Service and repair history	Warranty information
Vehicle handbook	Technical data manuals
Workshop manuals/Wiring diagrams	Safety recall sheets
Manufacturer specific information	Information bulletins
Technical helplines	Advice from other technicians/colleagues
Internet	Parts suppliers/catalogues
Jobcards	Diagnostic trouble codes
Oscilloscope waveforms	On vehicle warning labels/stickers
On vehicle displays	Reference/Textbooks

Always compare the results of any inspection, testing or diagnosis to suitable sources of data. Remember that no matter which information or data source you use, it is important to evaluate how useful and reliable it will be to your safety, diagnostic, maintenance and repair routine.

Automotive Actuators & Waveform Analysis

Where to start?

Step 1
- A good systematic diagnostic routine should always begin with careful questioning of the driver to gather as much information about the symptoms and history of the fault as possible.

Step 2
- After a brief visual inspection for obvious signs of damage or safety issues, the system should be tested to try and recreate the fault.

Step 3
- The vehicle should then be scanned for diagnostic trouble codes and any codes should be recorded. (If possible, a full scan should be conducted, as issues in unrelated systems can sometimes affect the operation of others).

Step 4
- Any codes should be cleared and the vehicle should be tested over a complete drive cycle.

Step 5
- Rescan the vehicle and concentrate diagnosis around any codes that have returned.

Step 6
- Connect the oscilloscope to the suspected circuit and analyse any waveforms produced.

Actuators (output)

Actuators provide a physical change to a vehicle system when signalled by the ECU. They are often types of electric motor (rotating or solenoid), but could produce heat, radio waves or other types of action.

Introduction

It is worthwhile remembering that a systematic diagnostic routine can help reduce wasted time, excessive effort and frustration. As a result, it is always good practice to start any fault diagnosis at the end. As strange as that might sound, if you think about it, all systems are designed to produce a result and if that result is correct, then operation happens as expected. Checking signals at the actuating components will tell you if commands are correctly getting through to their destination, or if there is any miscommunication on the way. If the signal has reached the actuator component successfully and in the correct format, then the likely-hood is that it's the component itself that is at fault. However, careful analysis of the waveform should always be conducted as a great deal of information can be obtained which can help with a successful diagnosis leading to a first-time fix.

Automotive Actuators
& Waveform Analysis

Examples of actuator types and a brief description of their purpose is shown in table 3.2.

Table 3.2 Examples of actuators and their purpose

Actuator	Description
EVAP purge valve 	The EVAP system is a method of safely storing fuel vapour fumes created in the fuel tank, and then releasing them through the combustion process as a method of emission control. Often referred to as the 'charcoal canister' the evaporative emissions system will contain a solenoid actuator. When engine operating conditions meet the prescribed limits, it will open and allow the fumes to be drawn into the intake manifold and be burned during the normal combustion process. This has no effect on vehicle performance, but does prevent the release of hydrocarbons to atmosphere.
Glow plugs 	Diesel engines use glow plugs as a method of improving cold starting, and on modern vehicles, as a method of lowering cold start emissions. Mounted in the combustion chamber, the glow plug is an electrically heated element which helps raise the temperature of the incoming air charge. A glow plug light is often incorporated on the driver's dashboard display which gives an indication of the waiting time needed before cold starting should be attempted. In many cases, this warning light does not give an indication of the actual operation of the glow plugs, but is simply connected to a separate timer circuit. Many glow plugs will actually run for a set period of time after the light has extinguished in order to help reduce cold start emissions (after glow).
EGR valve 	Exhaust gas recirculation EGR is a method of reintroducing a small amount of exhaust back into the intake of an engine which helps reduce emissions. The recirculated exhaust gas enters the cylinder with the induced air, taking up space that would normally be used for the fresh air/fuel charge and slightly lowering performance. This reduced cylinder performance will lower the combustion chamber temperatures and therefore the amount of NOx produced. An EGR valve is required so that exhaust gas is only recirculated when outright performance is not required. The engine is monitored by the ECU and a **solenoid** valve is actuated to control when and how much EGR is used.
Fuel pump 	The electric fuel pump on many petrol and Diesel vehicles is submerged in the fuel tank to help cool, silence and quench any sparks produced by the motor. Attached to the end of the electric motor is a roller-cell pumping mechanism that draws in fuel and supplies it to the rest of the system at a predetermined delivery pressure.

Automotive Actuators
& Waveform Analysis

Table 3.2 Examples of actuators and their purpose

Actuator	Description
IAC valve 	Idle speed (tick-over) of a petrol engine is determined by the amount of air allowed to enter the engine cylinder. Increase the air and the engine speed will increase, reduce the air and the engine speed will fall. The component that mainly controls the amount of air entering an engine is the throttle butterfly which will have an initial idle setting to maintain vehicle tick-over. When load is placed on the engine by systems such as power assisted steering or electrical demand via the alternator, it's not unusual for the idle speed to drop to a level where the engine may stall. To help overcome this issue, petrol engines are fitted with an idle air control valve, IAC. The IAC is a motor operated valve which will allow a small amount of air to bypass the throttle butterfly when a control signal is sent from the ECU. The motor is either a rotational type or a solenoid and the operation of the valve is regulated by **duty cycle**.
Petrol injector 	A petrol fuel injector is a solenoid operated valve, which when actuated by the ECU will move an internal **armature**, lifting a **pintle** needle off its seat and allow pressurised fuel to be injected into the engine. When it is switched off, a return spring closes the pintle and injection stops. The amount of time the injector is held open dictates the amount of fuel that is injected, and this is controlled from the ECU by **pulse width modulation PWM**.
Diesel injectors (solenoid and piezo electric) 	Modern Diesel injectors come in two main forms, solenoid and **piezoelectric**. Mounted directly in the combustion chamber and connected to a common high pressure source, when activated they will open and allow fuel to be injected at extremely high pressures. Solenoid injectors work in a similar manner to their petrol counterparts, in that, the armature is drawn open by the electromagnetic field created by the solenoid winding. Piezoelectric injectors on the other hand work in a different way. A stack of wafer thin crystals is mounted above the needle mechanism, and when an electric current is passed over them they deform and expand, opening the injector needle with extreme speed. Unlike the solenoid type, however, it is not simply closed by a return spring, but the **polarity** must be reversed to allow the piezo crystals to contract and close the injector.
Pressure regulator 	Accurate fuel pressure control is vital in modern injection to maintain precise timing and quantity measurement for injection purposes. Early fuel injection systems used a mechanical pressure regulator, but many vehicles now rely on electronic pressure measurement and control, especially in common rail Diesel, gasoline direct injection and return-less systems. The purpose of a pressure regulator is to maintain fuel injection pressures within very tight limits.

Automotive Actuators & Waveform Analysis

Table 3.2 Examples of actuators and their purpose

Actuator	Description
Quantity control valve 	The quantity control valve can be found on common rail Diesel fuel systems. Its purpose is to regulate the amount of fuel flowing in the low-pressure circuit between the feed pump in the tank and the high pressure engine driven pump. Regulating this flow can help maintain pressure in the common rail between set limits. (More fuel results in higher pressures in the rail and less fuel results in lower pressures). If the quantity of fuel in the low-pressure circuit is controlled, this reduces load on the high-pressure pump and helps keep fuel temperatures down.
Throttle servo motor 	Accurate engine management, safety and comfort/convenience systems have led to the demise of the mechanical throttle cable. This physical component has now been replaced by a drive-by-wire system on many cars. Working in conjunction with an accelerator pedal position sensor, the throttle **servo** motor actuates the throttle butterfly to regulate the air entering the engine. By controlling the throttle in this manner, other vehicle systems can also have an input into the vehicle load control. These systems may include: traction control, **ESP**, cruise control, active braking, automatic and semi-automatic transmission systems and many more.
Cooling fan 	Maintaining operating temperatures between set limits is vital to correct function and operation of the vehicle's engine. Too hot or too cold may have serious consequences for the health and lifespan of an engine. Also, if engine temperature is kept at the recommended manufacturer values, fuel economy, exhaust emissions and performance are maximised. The engine radiator cooling fan is an electric motor, which with ECU duty cycle control can now be regulated for speed and therefore air flow. Other vehicle cooling fans work in a similar manner, for example: air conditioning condenser fans and hybrid electric battery cooling fans.
VVT control 	Variable valve technology is now a common feature within engine design. It is a method that allows manufacturers and designers to enhance the performance of engines by increasing the **volumetric efficiency** over a larger rev range. The two main methods used in variable valve control are phasing (where the valve timing is advanced or retarded while the engine is running) and lift (where the amount that the valve is opened is varied while the engine is running). Most variable valve systems are controlled through engine oil hydraulic operation. Engine oil under pressure is directed to the appropriate chamber or piston within the VVT mechanism and the flow is controlled by some form of solenoid valve. The control of the solenoid is achieved by duty cycle.

Automotive Actuators & Waveform Analysis

Table 3.2 Examples of actuators and their purpose

Actuator	Description
Starter motor 	The starter motor connects to the engine crankshaft via the ring gear around the edge of the flywheel. Its purpose is to rotate the engine crankshaft at a speed that will allow the engine to ignite the fuel and start. (This is often in excess of 200 rpm). The motor should be small and compact, but still be able to deliver the turning effort and speed required for starting. The large amount of torque required is achieved by increasing the gear ratio between the starter pinion gear and the flywheel ring gear. Although this method of torque multiplication has been used successfully for many years, the trend towards smaller more powerful starters has led some manufactures to include an internal gearing mechanism in the motor design.
Alternator 	The alternator is an engine driven generator used to keep the vehicle battery charged and provide an auxiliary power source for the cars on-board electrics. It works by rotating an electromagnet inside a copper winding to induce an alternating current, which can then be converted into direct current by a **rectifier** before charging the battery. Most alternators have three copper windings to increase the output and this is known as 'three phase'. As the alternator spins faster, the output increases, and voltage has to be regulated to prevent damage to the battery and other electrical systems. Some manufacturers use smart charging technology, where the state of charge (SOC) of the battery is monitored and the alternator is allowed to temporarily overcharge or switch off altogether, depending on system requirements.
ABS solenoid 	The ABS **modulator** unit contains a series of hydraulic valves, which can control the fluid pressure in the braking system, to regulate and try to prevent wheel lock-up in hard braking situations. Solenoids are connected to these valve units and when modulated by duty cycle, are able to give three positions of operation: Closed Partially open Fully open This allows the hydraulic system to increase pressure, hold pressure and release pressure many times a second during an emergency braking situation.

Solenoid – a form of linear motor that moves backwards and forwards rather than rotating.

Duty cycle – a method of electronic control where components are rapidly switched on and off.

Armature – the central shaft of a motor.

Automotive Actuators & Waveform Analysis

Pintle – a small pin like valve found within some fuel injectors.

Pulse width modulation PWM – a method of electronic control where components are switched on for a varying amount of time.

Piezoelectric – the ability of certain materials to generate electricity due to mechanical stress or vibration.

Polarity – the direction of connection to an electric circuit (i.e. positive or negative).

Servo – a component that increases mechanical effort.

ESP – Electronic Stability Program.

Volumetric efficiency – how well an engine cylinder can be filled with incoming air.

Rectifier – the component in an alternator that converts alternating current (AC) to direct current (DC).

Modulator – a component that varies a process or action.

Solenoids

A solenoid is a linear motor, which when activated, will cause the armature to slide sideways within an electromagnet. Many solenoids are returned to their original starting position by a spring when de-energised.

Rotating Motors

Unlike a solenoid, a standard motor is designed to rotate. When energised, an electromagnetic field created in a copper conductive winding will attract or repel against a permanent fixed magnet, making the motor turn. Once out of the magnetic field, the polarity of the electromagnet must be reversed to keep the armature turning; this is achieved using a **commutator**.

Commutator – a segmented electrical contact inside an electric motor, designed to swap polarity as the armature rotates.

An important point to remember when testing actuators is that many of them may be based on motors (either rotational or linear solenoid types), therefore a crossover may be observed in the waveform patterns produced and many will look similar. It is important to understand how the component functions in order to distinguish any differences and form a satisfactory conclusion on its operation.

Automotive Actuators & Waveform Analysis

Oscilloscope quick set up guide for use with actuators:

Measuring voltage:

How to:

Note: The oscilloscope probes may come in different colours, but for the sake of simplicity we will call them red and black here.

Step 1
- Connect the tip of the black lead to a good source of earth, such as the battery terminal, metal bodywork or engine. This will then only leave you with the red wire to worry about.

Step 2
- Now connect the red probe to the circuit to be tested.

Step 3
- Adjust the scales until you see an image on the screen.

Step 4
- After some practice, you will become familiar with the patterns and waveforms created by different vehicle systems.

Some actuators will produce a high voltage spike as they switch off, known as a 'back EMF'. This voltage spike may require the use of an attenuator, in order to correctly read and interpret the waveform shown on the display.

EVAP purge valve

The EVAP system, or charcoal canister as it is known by many technicians, is a method of storing the fumes produced in the fuel tank and releasing them into the engine intake system so that they can be burnt during the engines normal combustion cycle. This emission control method will actuate when the engine has reached its normal operating temperature, and a solenoid valve will open allowing the fumes to be released from the charcoal canister into the manifold.
To test the solenoid, voltage probes should be attached to the purge valve.

Automotive Actuators & Waveform Analysis

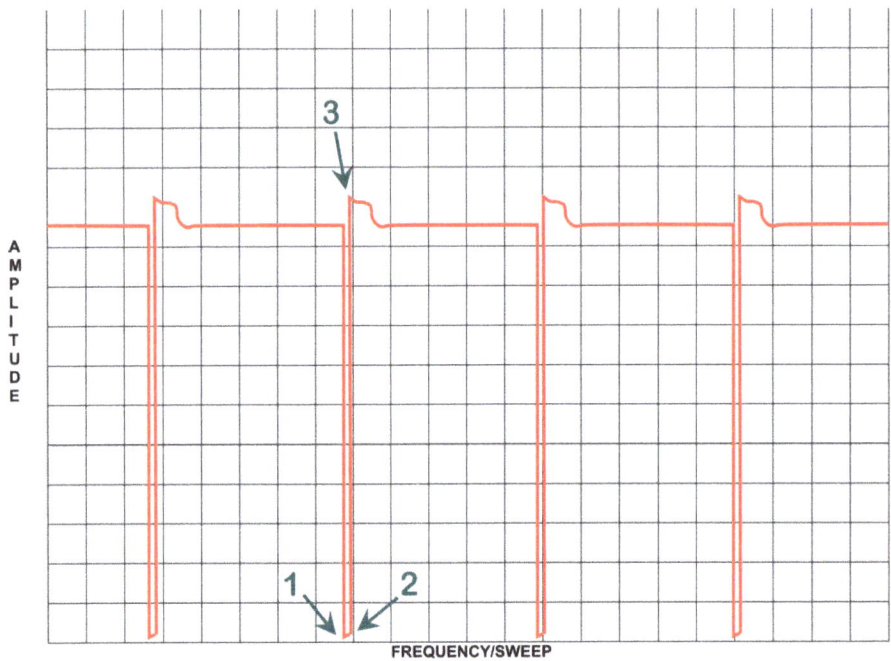

Figure 3.3 EVAP (charcoal canister) purge valve waveform

Table 3.3 Waveform analysis EVAP solenoid valve

Waveform component	Description
1	This is the point on the waveform that the EVAP solenoid valve is switched on/opened by the ECU. It is switched to earth, so the voltage will fall very close to 0 volts.
2	This is the point on the waveform that the EVAP solenoid is switched off/closed by the ECU. When current is removed, a spring inside the valve returns it to the closed position.
3	This spike in the waveform is caused by the collapsing magnetic field within the solenoid winding and is an induced voltage that will appear on some solenoids when switched off.

Remember that the engine needs to have reached its normal operating temperature before the EVAP solenoid will be activated. Conducting a waveform analysis before this may lead to misdiagnosis.

Automotive Actuators & Waveform Analysis

Glow plugs

Glow plugs are used on Diesel engines to pre-heat the air in the combustion chamber for cold start and emission purposes.
Information on correct operation can be gained by attaching an inductive amp adapter to your oscilloscope and connecting it to the feed wire that leads to the glow plugs.

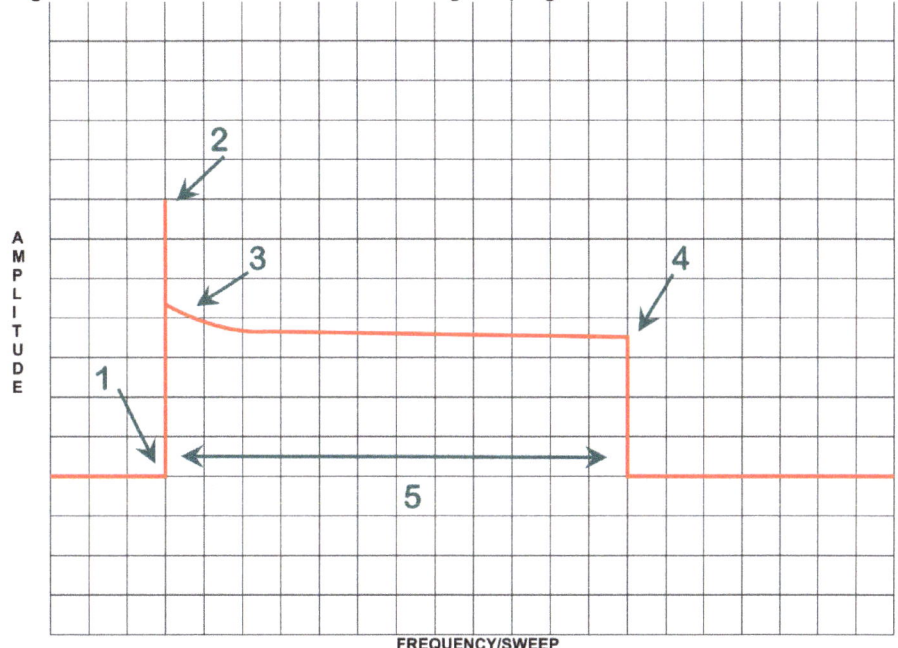

Glow plugs

Connect to the circuit using:

Inductive Amps Clamp

Figure 3.4 Glow plug current waveform

Table 3.4 Waveform analysis glow plug current

Waveform component	Description
1	This is the point on the waveform that the glow plugs are switched on and current starts to flow, causing the waveform to rise rapidly.
2	This is the point on the waveform where the initial current draw peaks, showing the maximum switched on current.
3	Once the current starts to flow, and the glow plug heats up, the amperage will fall and settle to a relatively steady pattern.
4	This is the point on the waveform that the glow plugs are switched off by the timer relay circuit. When power is removed, current stops and the waveform returns to zero.
5	The distance between the on and off components of the waveform can give a very accurate illustration of the period of time the glow plugs are switched on.

Automotive Actuators & Waveform Analysis

Tip

An indication of the expected current draw for the glow plugs can be calculated using the following formula:
Wattage of one glow plug, multiplied by the number of cylinders, divided by battery voltage.
i.e. Current draw (in amps) = wattage ÷ voltage

Some Diesel engines modulate and control the operation of the glow plugs for a longer period of time in order to help reduce emissions during cold starting and warm-up. This can be achieved by regulating or limiting the current draw via a control module.

Information on correct operation can be gained by attaching an inductive amp adapter to your oscilloscope and connecting it to the feed wire that leads to the glow plugs.

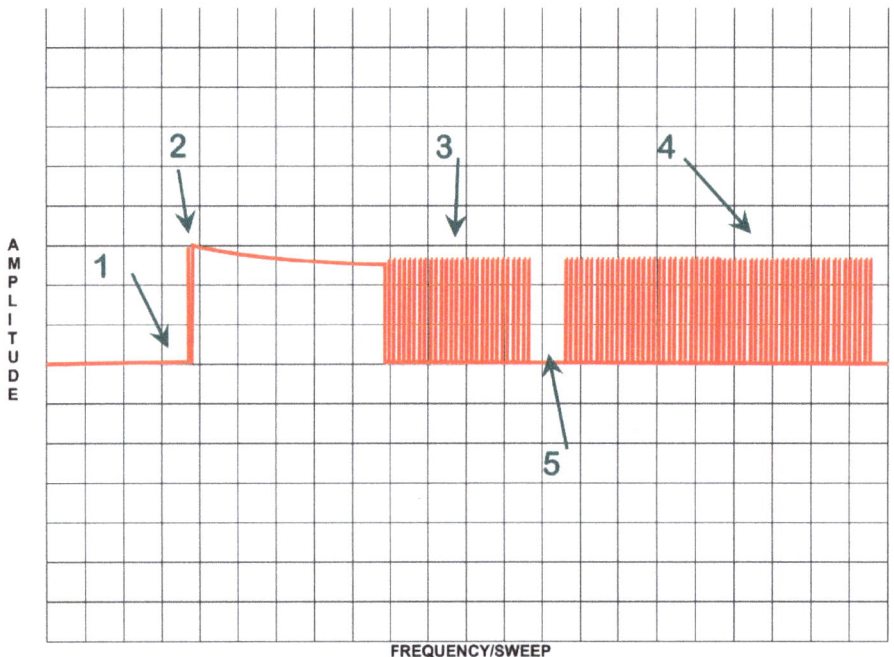

Figure 3.5 Glow plug current (modulated) waveform

Table 3.5 Waveform analysis glow plug current (modulated)

Waveform component	Description
1	This is the point on the waveform that the glow plugs are switched on and current starts to flow, causing the waveform to rise rapidly.
2	This is the point on the waveform where the initial current draw peaks, showing the maximum switched on current.
3	Once the current starts to flow, and the glow plug heats up, the amperage will fall and then be regulated using a control module.
4	This point on the waveform shows a regulated current supply which is known as 'post-heat' and helps control warm-up emissions.
5	During cranking, current to the glow plugs is switched off to reduce excessive electrical load on the vehicle systems.

Automotive Actuators & Waveform Analysis

EGR valve

The EGR (exhaust gas recirculation) valve is a solenoid device which allows small amounts of exhaust gas to re-enter the engine to reduce combustion temperatures and the production of the pollutant NOx (oxides of nitrogen). The valve may be directly connected to one of the manifolds, or remotely through a vacuum valve.
The control of the EGR valve is achieved through duty cycle, where rapid switching on and off will regulate the opening of the solenoid. The 'on time' is a percentage of the total on/off period of one switched cycle.
To test the solenoid, voltage probes should be attached to the EGR valve.

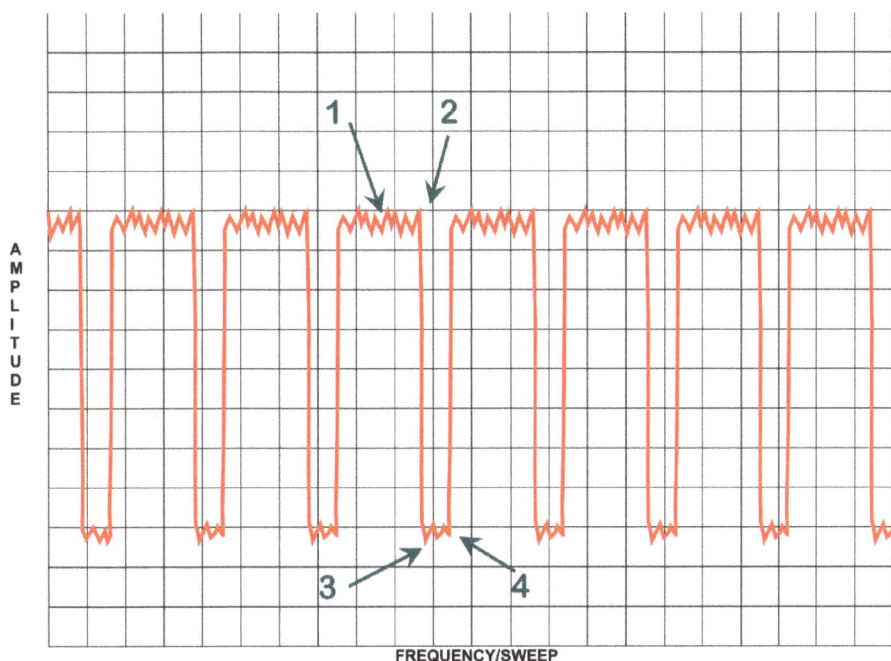

Figure 3.6 EGR valve duty cycle waveform

Table 3.6 Waveform analysis EGR valve

Waveform component	Description
1	This is the point on the waveform which shows the switched-off period and is measured as a percentage of the total on/off time. It is not unusual to have small amounts of interference on the waveform at this point.
2	This is the point on the waveform where the EGR valve is switched-on and opens.
3	This is the point on the waveform which shows the switched-on period and is measured as a percentage of the total on/off time. It is not unusual to have small amounts of interference on the waveform at this point.
4	This is the point on the waveform where the EGR valve is switched-off and closes. (It not unusual to have a small voltage spike at the top edge of the switch off point on the waveform caused by the collapsing magnetic field inside the solenoid).

Automotive Actuators & Waveform Analysis

EGR only takes place under specific conditions which include, vehicle speed, engine temperature and engine load.
To correctly measure EGR operation it may be necessary to road test the vehicle in order to achieve a waveform.

To safely take oscilloscope readings while road testing, you will need to perform the diagnosis with the aid of an assistant, or place the scope where it will not affect safe driving, and use a flight record facility.

Fuel pump

The electric fuel pump of many vehicles is submerged in the fuel tank to assist with cooling, silencing, suppression of sparks (and therefore sources of ignition). Normal diagnosis and testing of this component can be very involved and time consuming.

An oscilloscope and inductive amp clamp connected around the feed wire to the pump, or a fused loop inserted instead of the fuel pump fuse, can give easy access to information about the operation of a running pump with very little stripping down. (This also gives the opportunity to test the component in-situ, providing a realistic operating condition to aid your diagnosis).

When the fuel pump is running, a distinctive ripple waveform is produced by the motors commutator, which can then be examined for bad segments.

Many fuel pumps have six to eight commutator segments, so any unusual patterns that repeat with this frequency can be diagnosed as poor or failing commutator sections.

Most fuel pumps will have a running current of around three to five amps (with the exception of early Bosch K Jetroninc which has around eight amps).

This amperage can help determine the operating condition of the pump:
- Too low – the pump is finding it too easy (possible internal leak in the fuel system).
- Too high – the pump is struggling (possible internal restriction or blockage in the fuel system).

Automotive Actuators & Waveform Analysis

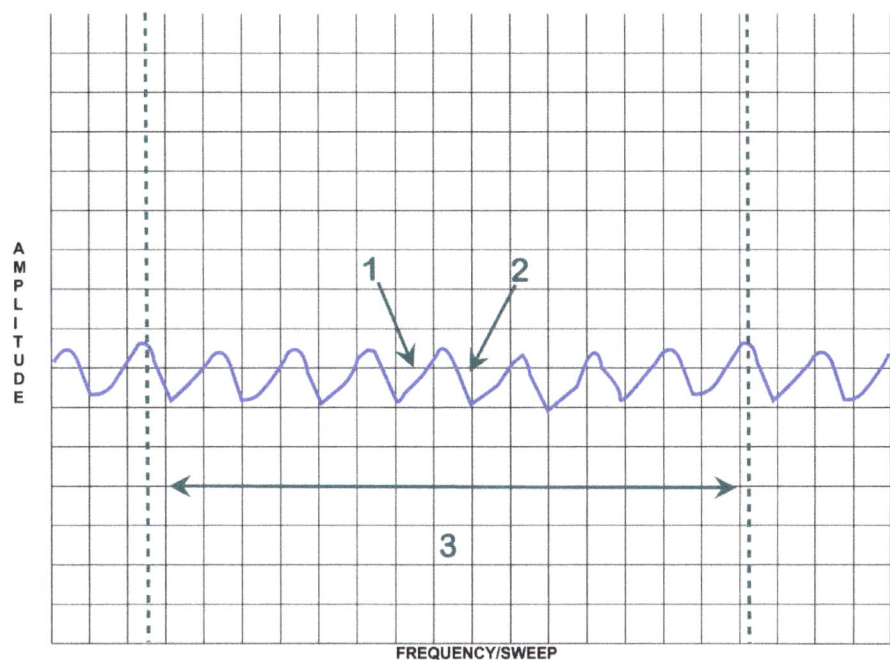

Figure 3.7 Electric fuel pump commutator waveform

Table 3.7 Waveform analysis fuel pump current

Waveform component	Description
1	This is the point on the waveform which shows a section of commutator aligning with a brush and current flow building in the armature electromagnet coils. (i.e. powering up).
2	This is the point on the waveform which shows a section of commutator moving away from the brush and current flow falling in the armature electromagnet coils. (i.e. powering down).
3	The distance between the two cursors added to this waveform gives an example of one complete revolution of a motor with eight commutator sections. Some oscilloscopes will display the time period (Delta Δ) between the cursors which can help you calculate the rotational speed of the pump.

To calculate the RPM of the fuel pump from the current ripple if the timescale is set to milliseconds (m/s):
Add cursors to the display and position on repeating points of the waveform pattern (i.e. six or eight ripples apart).
Take the Delta Δ time displayed and perform the following calculation:
60,000 ÷ Delta Δ = fuel pump RPM
An average petrol fuel pump will have a rotational speed of between 3,000 to 7,000 RPM.

Automotive Actuators & Waveform Analysis

IAC valve – solenoid and rotary types

The idle air control valve (IAC) helps to maintain tick-over when loads are placed on the engine, which might normally lead to stalling. It does this by operating a valve that allows a small amount of air to bypass the throttle butterfly and is regulated by the engine ECU.
The control of the IAC valve is achieved through duty cycle, where rapid switching on and off will regulate how far the valve opens. The on time is a percentage of the total on/off period of one switched cycle.
To test the valve, voltage probes should be attached to the IAC.

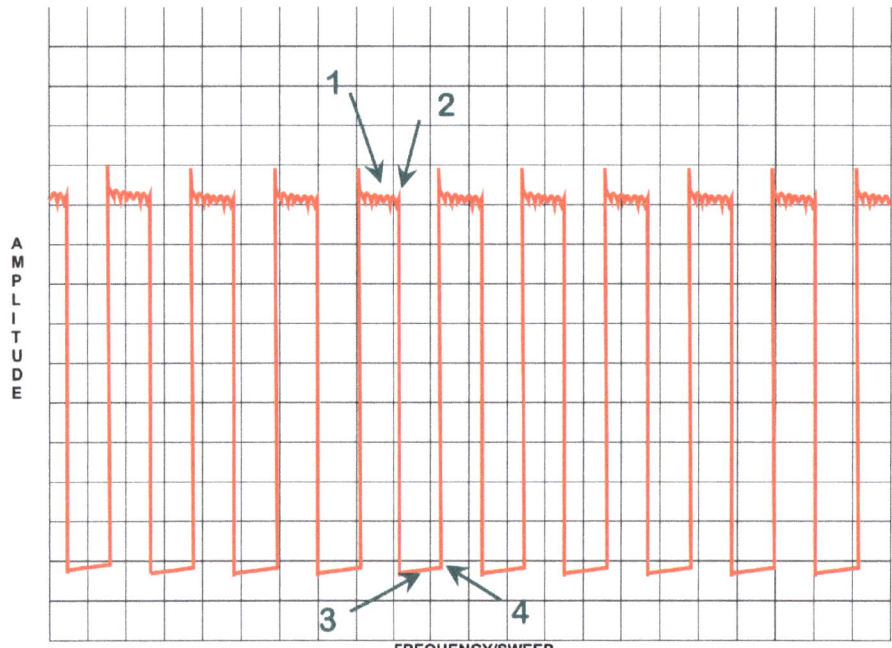

Figure 3.8 Idle air control valve waveform

Table 3.8 Waveform analysis IAC valve rotary and solenoid

Waveform component	Description
1	This is the point on the waveform which shows the switched-off period and is measured as a percentage of the total on/off time. It is not unusual to have small amounts of interference on the waveform at this point.
2	This is the point on the waveform where the IAC valve is switched-on and opens.
3	This is the point on the waveform which shows the switched-on period and is measured as a percentage of the total on/off time. It is not unusual to have small amounts of interference on the waveform at this point.
4	This is the point on the waveform where the IAC valve is switched-off and tries to close. (It not unusual to have a small voltage spike at the top of the switch off point on the waveform caused by the collapsing magnetic field inside a solenoid).

Automotive Actuators & Waveform Analysis

> Many idle air control valves use two wires, one powered and one switched earth through duty cycle.
> If the IAC uses three wires, it has two switched earths; one to open and one to close.
> Use a two-channel set-up on the oscilloscope to compare the waveforms from both wires.

Stepper motor IAC

The idle air control stepper motor helps to maintain tickover when loads are placed on the engine which might normally lead to stalling. It does this by operating a motor in gradual steps, allowing a small amount of air to bypass the throttle butterfly, or operates against the throttle mechanism, and is regulated by the engine ECU.
Rarely used on modern engines, two main types are available; four and five wire.

A four-wire stepper motor is switched both positive and negative when the throttle is signalled to be in the closed position.

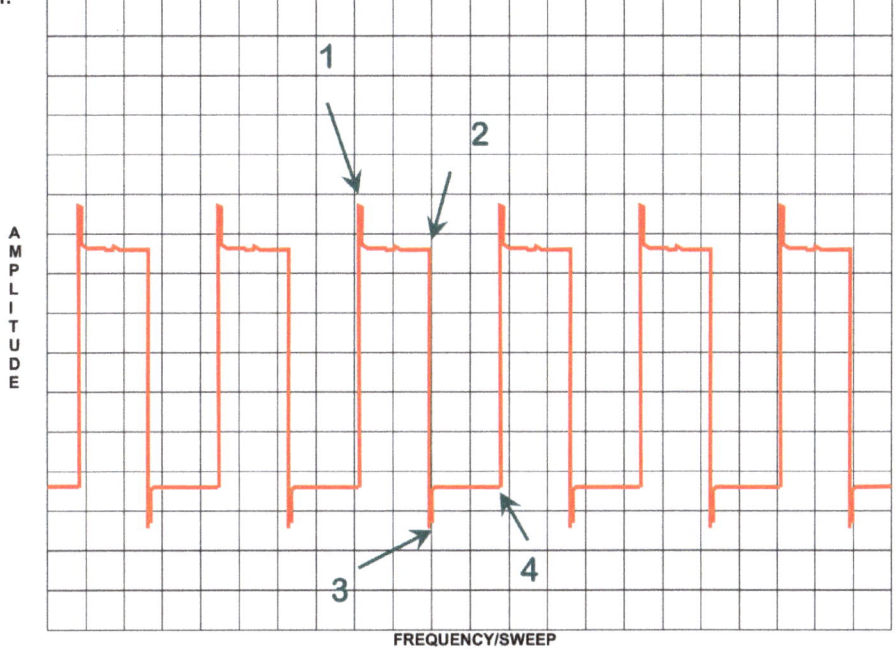

Stepper motor IAC

Connect to the circuit using:

Voltage Probes or Clips

Figure 3.9 Idle air control 4 wire stepper motor waveform

Table 3.9 Waveform analysis IAC 4 wire stepper motor

Waveform component	Description
1 & 2	These are the points on the waveform showing the stepper motor being switched to positive. (It's not unusual to have a small voltage spike at the top of the switched positive on the waveform, caused by the collapsing magnetic field inside the solenoid).

Automotive Actuators & Waveform Analysis

Table 3.9 Waveform analysis IAC 4 wire stepper motor

3 & 4	These are the points on the waveform showing the stepper motor being switched to negative. (It's not unusual to have a small voltage spike at the bottom of the switched negative on the waveform, caused by the collapsing magnetic field inside the solenoid).

A five-wire stepper motor has four switched earths, allowing distinct opening positions which can be seen and compared when using four channels on the oscilloscope.

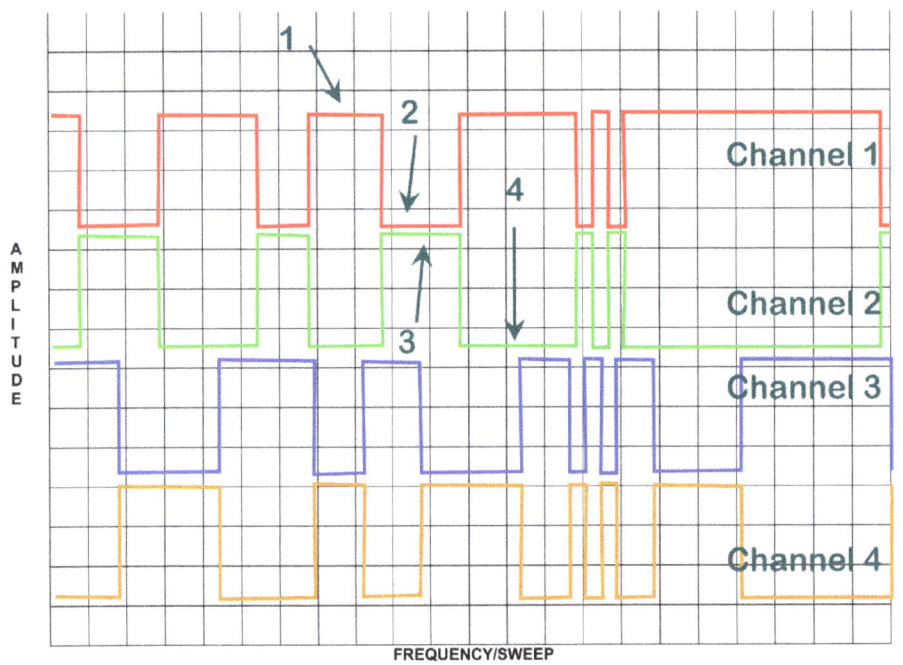

Figure 3.10 Idle air control 5 wire stepper motor waveform

Table 3.10 Waveform analysis IAC 5 wire stepper motor

Waveform component	Description
1	This is the point on the waveform which shows the stepper motor being switched on for channel 1.
2	This is the point on the waveform which shows the stepper motor being switched off for channel 1.
3	This is the point on the waveform which shows the stepper motor being switched on for channel 2.
4	This is the point on the waveform which shows the stepper motor being switched off for channel 2.
Note	The same switching occurs for both channels 3 and 4.

Automotive Actuators & Waveform Analysis

Petrol injectors

Multipoint petrol fuel injectors are most commonly small solenoid operated valves which allow pressurised fuel to be injected into the inlet manifold, just before the inlet valve.
There are two main types of solenoid used, high resistance (8 ohms to 13 ohms) or low resistance (0.3 ohms to 0.8 ohms) approximately.

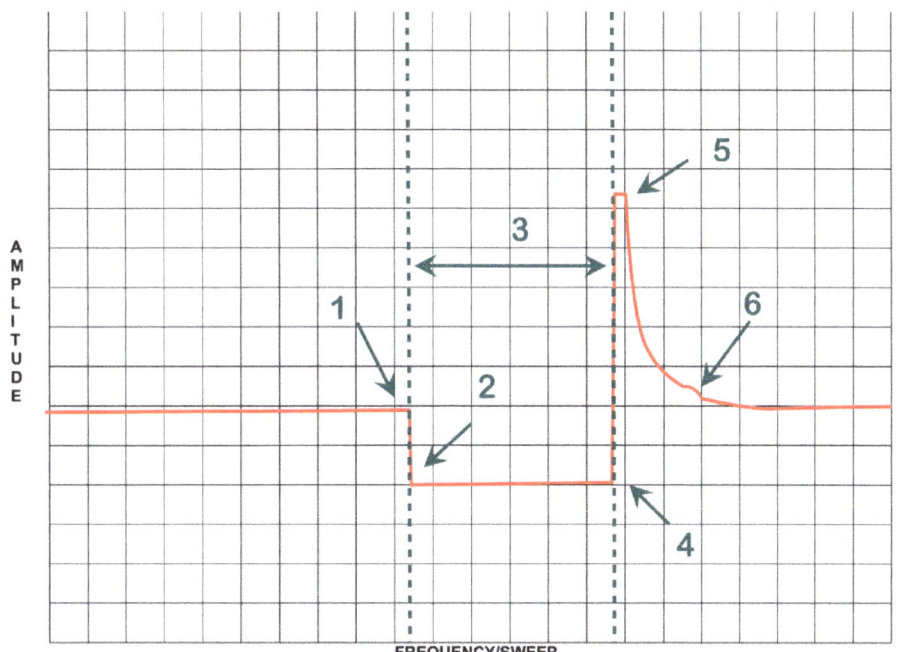
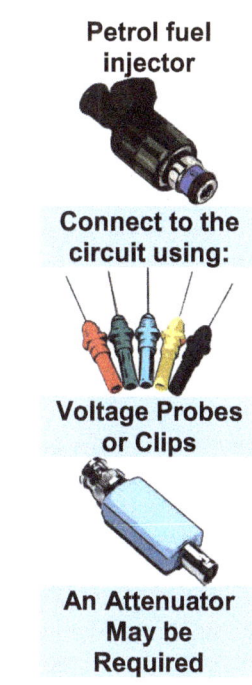

Figure 3.11 Multipoint injector voltage waveform

Table 3.11 Waveform analysis multipoint petrol injector voltage

Waveform component	Description
1	This is the point on the waveform which shows the injector being switched on by the ECU.
2	As it is switched to earth, the voltage will drop to nearly 0 volts.
3	The distance between the two cursors is the switch on time of the injector, known as the 'pulse width'.
4	This is the point on the waveform where the injector is switched off by the ECU.
5	The voltage spike is the result of the collapsing magnetic field created in the solenoid winding as it is switched off and is known as a 'back EMF'. This voltage spike needs to be limited on some styles of injector to prevent long-term damage to the injector needle and its seat. (The voltage spike can make the needle bounce against its seat as it closes). As a result, the spike is 'clipped' by a zenner diode and the top of the spike will be flattened off.
6	This small hump in the injector pattern is caused by the disruption in the magnetic field as the injector armature moves, and indicates the point at which the needle closes.

Automotive Actuators & Waveform Analysis

Tip: If the system uses a zenner diode and the spike is not squared off, this may be an indication of a weak winding in the coil of the injector solenoid.

If a zoom function is available on the oscilloscope, it may be possible to focus in on certain points and get further diagnostic information.

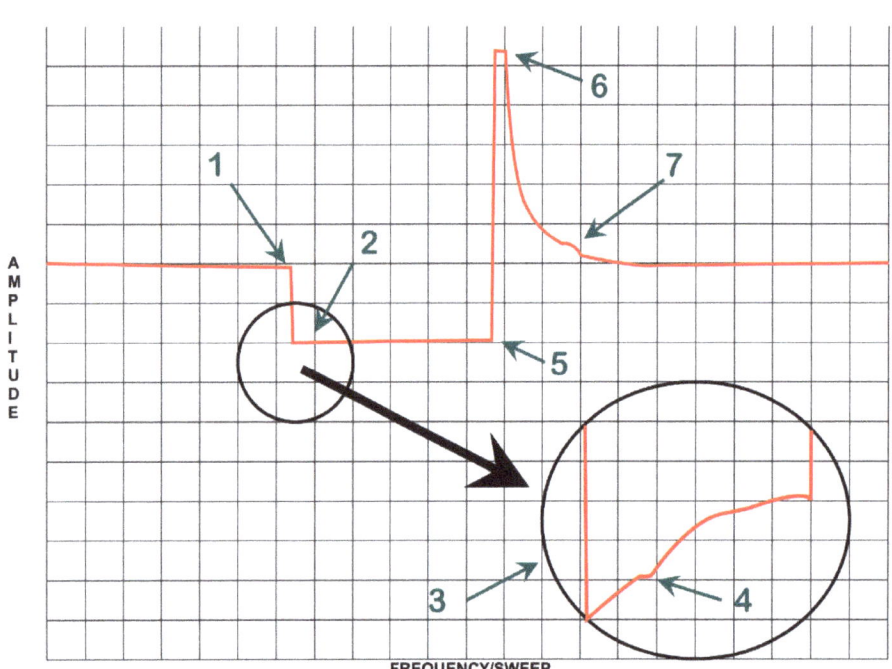

Figure 3.12 Multipoint injector voltage magnified waveform

Table 3.12 Waveform analysis multipoint petrol injector voltage magnified

Waveform component	Description
1	This is the point on the waveform which shows the injector being switched on by the ECU.
2	As it is switched to earth, the voltage will drop to nearly 0 volts.
3	In this magnified section, you can see a very small rise in voltage as the transistor switches the injector on and current starts to flow.
4	The small dip at this point in the magnified waveform is caused by the armature/needle opening.
5	This is the point on the waveform where the injector is being switched off by the ECU.

Automotive Actuators & Waveform Analysis

Table 3.12 Waveform analysis multipoint petrol injector voltage magnified

6	The voltage spike is the result of the collapsing magnetic field created in the solenoid winding as it is switched off and is known as a 'back EMF'. This voltage spike needs to be limited on some styles of injector to prevent long-term damage to the injector needle and its seat. (The voltage spike can make the needle bounce against its seat as it closes). As a result, the spike is 'clipped' by a zenner diode and the top of the spike will be flattened off.
7	This small hump in the injector pattern is caused by the disruption in the magnetic field as the injector armature moves, and indicates the point at which the needle closes.

A low resistance injector gives faster opening times, but would quickly overheat and burn out if the supply current was allowed to flow unregulated.

In a low resistance injector system, once the injector has been snapped open by the solenoid, current in the winding is modulated so that the valve remains open, but the winding is not overloaded and damaged.

The pattern created by the injector voltage will be similar to that of a high resistance type, however, during the 'on' period, a disruption to the waveform can be seen as the current is modulated.

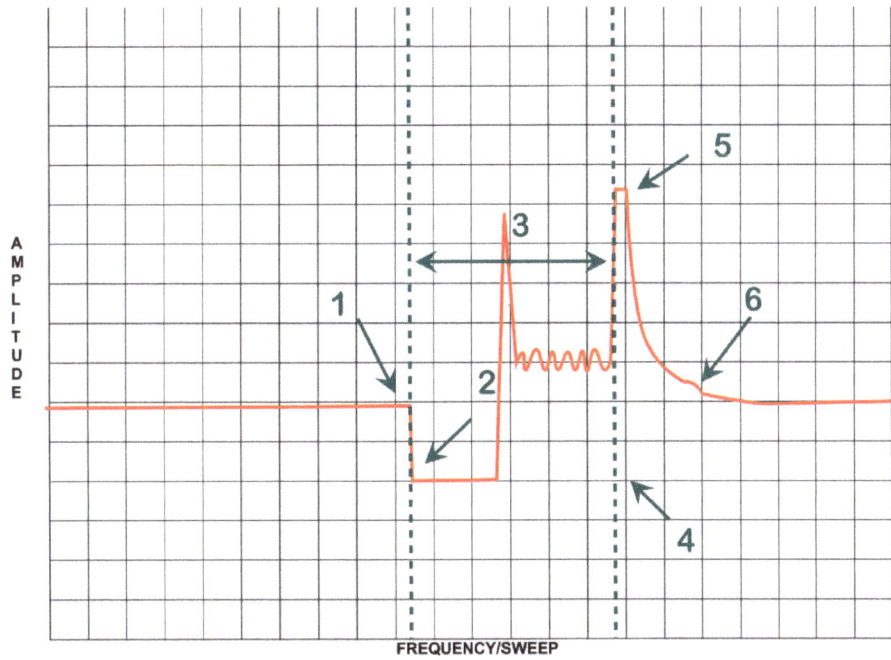

Figure 3.13 Low-resistance injector voltage waveform

Table 3.13 Waveform analysis low-resistance petrol injector voltage

Waveform component	Description
1	This is the point on the waveform which shows the injector being switched on by the ECU.

Automotive Actuators & Waveform Analysis

Table 3.13 Waveform analysis low-resistance petrol injector voltage

2	As it is switched to earth, the voltage will drop to nearly 0 volts.
3	The distance between the two cursors is the switch on time of the injector, known as the 'pulse width' and is made up of both the full current and modulated current sections.
4	This is the point on the waveform where the injector is switched off by the ECU.
5	The voltage spike is the result of the collapsing magnetic field created in the solenoid winding as it is switched off and is known as a 'back EMF'.
6	This small hump in the injector pattern is caused by the disruption in the magnetic field as the injector armature moves, and indicates the point at which the needle closes.

The operation of a petrol fuel injector can also be assessed using current flow. By placing an amp clamp around one of the injector wires and running the engine, a distinctive sloping waveform should be produced.

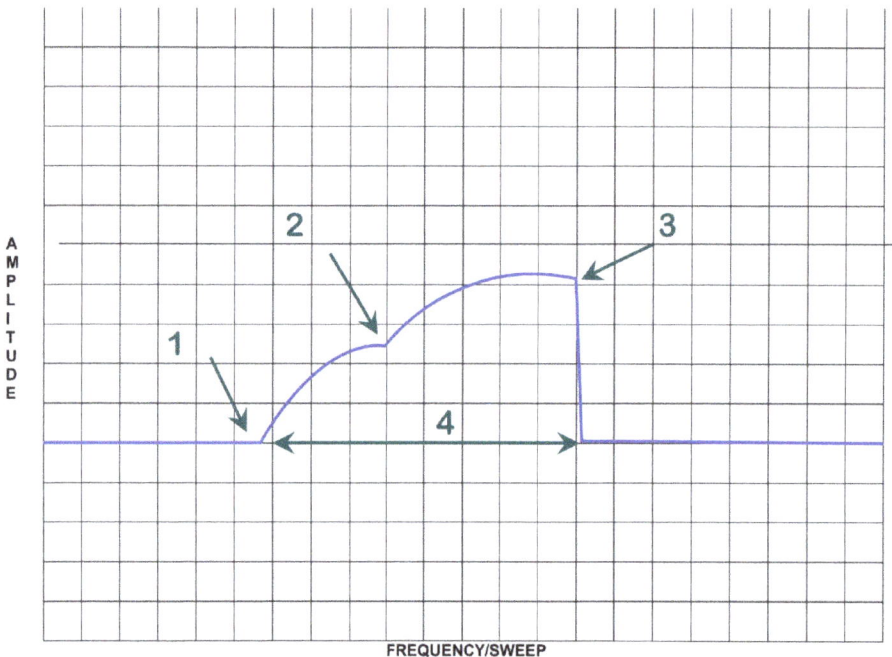

Figure 3.14 Multipoint injector current waveform

Table 3.14 Waveform analysis multipoint petrol injector current

Waveform component	Description
1	This is the point on the waveform that the injector is switched on and current starts to flow.
2	At this position, the electromagnetic field inside the solenoid has reached a point where it is strong enough to move the pintle and the injector opens. The opening of the pintle will cause a small disruption in the magnetic field and this is seen as a dip in the pattern here.

Automotive Actuators & Waveform Analysis

Table 3.14 Waveform analysis multipoint petrol injector current

3	The current flow in the solenoid winding continues to rise until the injector is switched off by the ECU.
4	The distance between these two points is the switched-on time of the injector, however, it should not be confused with the open time.

Tip

A good indication of the internal resistance of the injector can be gained by examining the rise of the current 'ramp'.
The waveform should rise in a steady curve and peak just before the injector is switched off.
If the current ramp rises too quickly, the internal resistance is too low.
If the current ramp rises too slowly, the internal resistance is too high.

If two channels are set up and used on your oscilloscope, then the voltage pulse width and the current ramp can be compared.

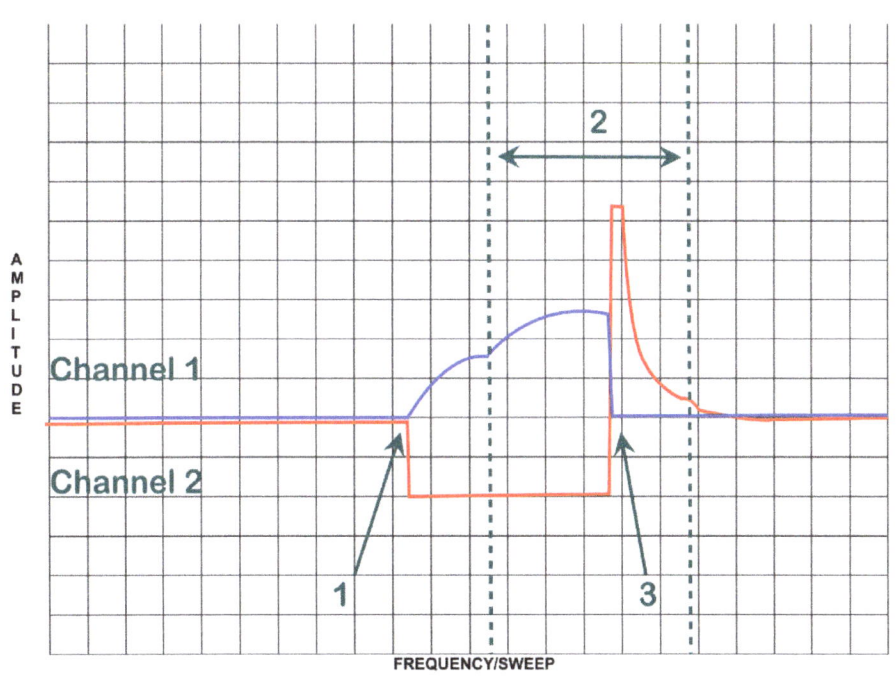

Figure 3.15 Multipoint injector voltage verses current waveform

Automotive Actuators & Waveform Analysis

Table 3.15 Waveform analysis multipoint petrol injector voltage verses current

Waveform component	Description
1	Channels 1 and 2. This is the point on the wave form where the ECU switches on the injector.
2	Channels 1 and 2. This shows cursors applied to the display and set at the points of magnetic field disruption on both the current ramp, and the falling voltage EMF, which represent the opening and closing of the injector needle. Some oscilloscopes will display the period (Delta Δ) between the cursors which can help you calculate the open time of the injector needle.
3	Channels 1 and 2. This is the point on the waveform where the ECU switches off the injector.

Single point injection

With single point fuel injection (sometimes known as throttle-body fuel injection) one injector is mounted in a similar position to a carburettor, just above a throttle butterfly. Fuel is injected as a fairly constant stream, and the amount is varied due to engine demands.

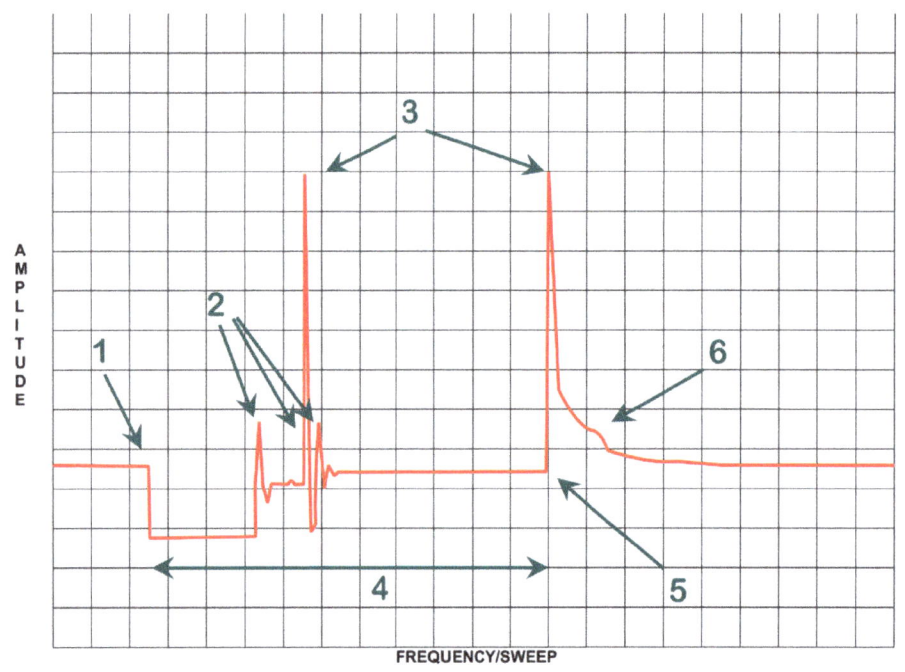
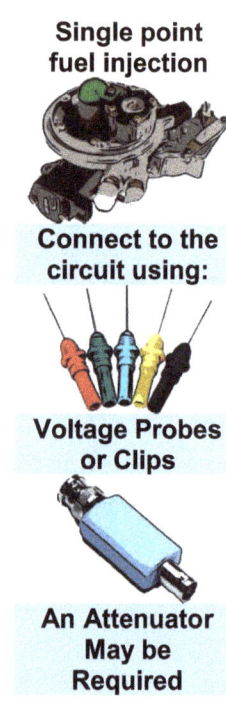

Figure 3.16 Single point injector voltage waveform

Automotive Actuators & Waveform Analysis

Table 3.16 Waveform analysis single point petrol injector voltage

Waveform component	Description
1	This is the point on the waveform where the injector is initially switched on.
2	The injector has multiple pulses at this point, helping to create the spray pattern.
3	These spikes on the waveform represent collapsing magnetic fields inside the solenoid injector. Depending on engine demands and speed, this period will expand and contract.
4	This is the overall injection switched on duration (pulse width).
5	This is the point on the waveform where the injector is switched off.
6	This small hump in the injector pattern is caused by the disruption in the magnetic field as the injector armature moves, and the point at which the needle closes.

On some larger engines, two fuel injectors may be used inside the same throttle body housing.

The operation of a single point petrol fuel injector can also be assessed using current flow. By placing an amp clamp around one of the injector wires and running the engine, a current flow waveform should be produced.

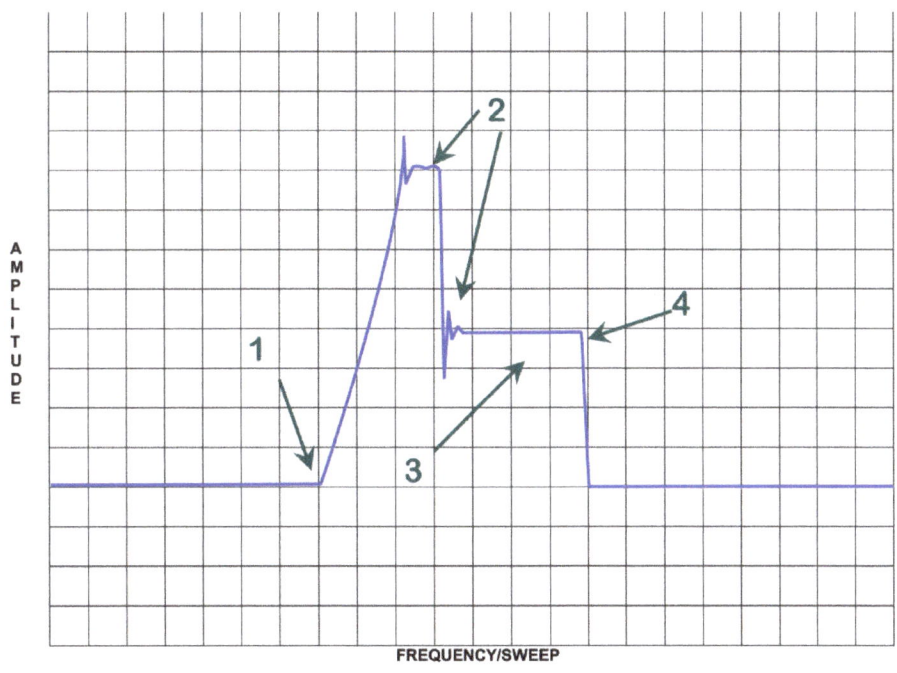

Single point fuel injection

Connect to the circuit using:

Inductive Amps Clamp

Figure 3.17 Single point injector current waveform

Automotive Actuators & Waveform Analysis

Table 3.17 Waveform analysis single point petrol injector current

Waveform component	Description
1	This is the point on the waveform where the injector is switched on and the current starts to flow.
2	This section is responsible for the magnetic field that opens the injector and may also represent multiple injector openings.
3	This shows a current limited section that is maintaining the injector in an open position, without overheating the solenoid.
4	This is the point on the waveform when the injector is switched off and the current flow stops.

Diesel injectors (solenoid and piezoelectric)

Modern high pressure common rail Diesel engines (CRD) use highly accurate electronic fuel injectors to accurately meter fuel injected into the combustion chamber.

The types used tend to be of two main design styles:
- Solenoid type
- Piezoelectric

Both injector styles have a rapid operating cycle and can be used to inject in a series of stages known as phases.

The phases consist of:
- Pilot – stage 1
- Main – stage 2
- Post – stage 3

Notes

The phased injection periods used on many high pressure common rail Diesel engines improve efficiency by creating a more controlled burn of the air/fuel mixture. This also leads to quieter running, lower emissions and better fuel economy, when compared to a standard Diesel engine.

Solenoid Diesel injector

A solenoid Diesel injector works in a similar manner to its petrol counterpart. When energised, the magnetic field created by a small coil of wire, draws an armature needle valve into an open position, and extremely high pressure fuel is injected into the combustion chamber. As soon as the injector is switched off, current flow stops and the injector needle is closed by a spring mechanism.

Due to their design and operation, most Diesel fuel injectors are best tested by connecting the oscilloscope to measure current, using an amp clamp.

When operated, solenoid Diesel injectors may produce slightly different waveforms depending on their design. The following section describes some waveform patterns that may be displayed.

Automotive Actuators & Waveform Analysis

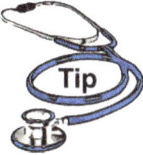

Tip: Many high pressure electronic Diesel injectors operate with voltages larger than battery/system voltage, and this must be considered when conducting waveform diagnosis.

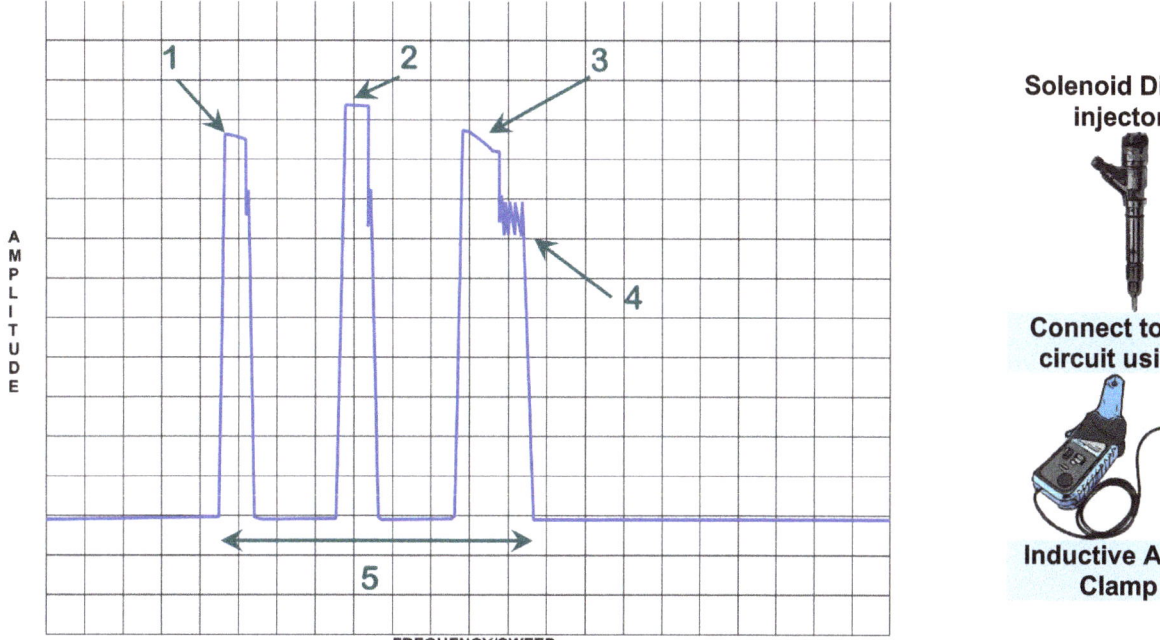

Figure 3.18 CRD solenoid injector current waveform version 1

Table 3.18 Waveform analysis CRD solenoid injector current version 1

Waveform component	Description
1	This section of the waveform represents the pilot injection, where a small amount of fuel is injected into the combustion chamber and starts the process of combustion. This provides a more controlled flame spread and helps to reduce the **delay period**.
2	This section of the waveform represents the main injection, which provides the majority of combustible fuel used in the production of the power stroke. Depending on engine loads and demand, this section of the waveform will expand and contract as the engine speeds up and slows down.
3	This section of the waveform represents the post injection period, which is used in some operating conditions to help reduce emissions. It is possible that this third injection phase may disappear completely under certain operating conditions.
4	This point on the waveform represents a small amount of current limiting as the injector is held open.
5	This period represents the total injection time for one cycle which includes all three phases of injection: pilot, main and post.

Automotive Actuators & Waveform Analysis

Delay period – the length of time between the injection of fuel and the start of combustion on a Diesel engine.

The following example waveform is also from a common rail solenoid type Diesel injector, however, due to its design the amplitude is lower and only pilot and main injection periods are shown.

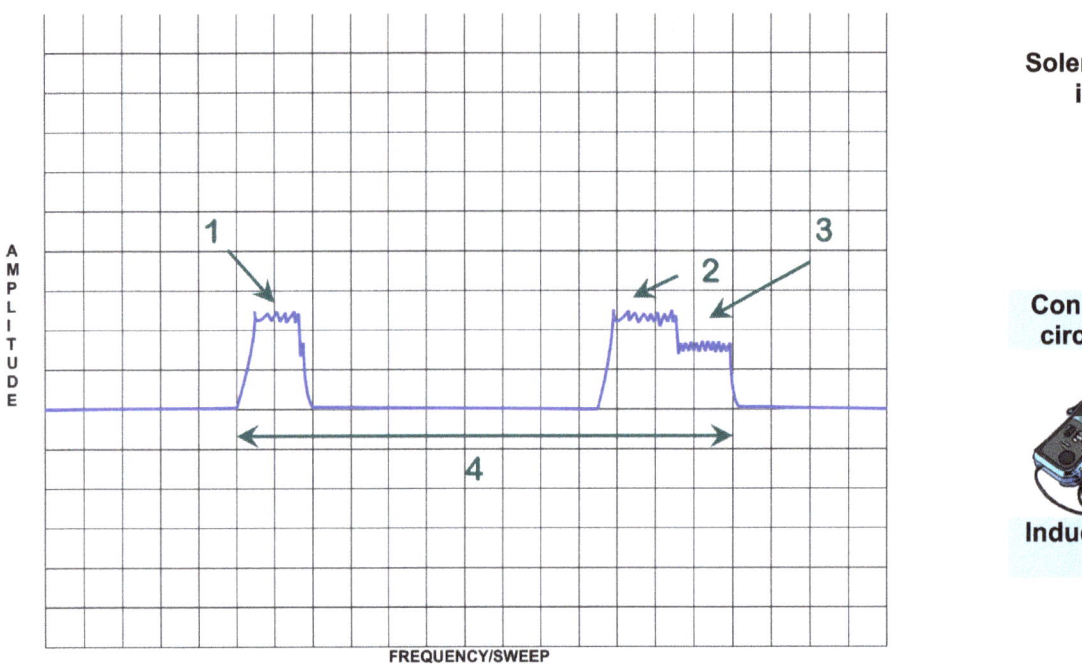

Figure 3.19 CRD solenoid injector current waveform version 2

Table 3.19 Waveform analysis CRD solenoid injector current version 2

Waveform component	Description
1	This section of the waveform represents the pilot injection, where a small amount of fuel is injected into the combustion chamber and starts the process of combustion. This provides a more controlled flame spread and helps to reduce the delay period.
2	This section of the waveform represents the main injection, which provides the majority of the combustible fuel used in the production of the power stroke. Depending on engine loads and demand, this section of the waveform will expand and contract as the engine speeds up and slows down.
3	This point on the waveform represents a small amount of current limiting as the injector is held open.
4	This period represents the total injection period for one cycle, but includes only two phases of injection: pilot and main.

Automotive Actuators & Waveform Analysis

The following image represents another style of solenoid injector waveform from a common rail Diesel engine, but due to design and operating conditions, this has two pilot injections followed by the main injection phase.

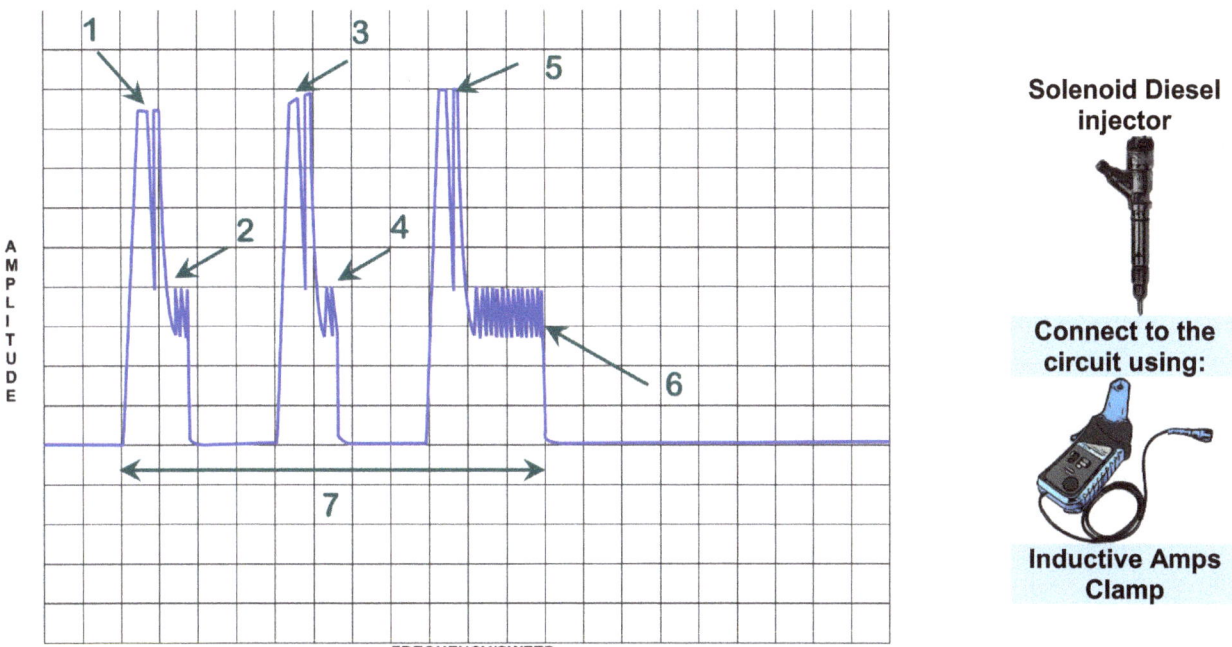

Figure 3.20 CRD solenoid injector current waveform version 3

Table 3.20 Waveform analysis CRD solenoid injector current version 3

Waveform component	Description
1	This section of the waveform represents the pilot injection, where a small amount of fuel is injected into the combustion chamber and starts the process of combustion.
2	This point on the waveform represents a small amount of current limiting as the injector is held open.
3	This section of the waveform represents a second pilot injection, where a small amount of fuel is injected into the combustion chamber providing a more controlled flame spread and helps to reduce the delay period.
4	This point on the waveform represents a small amount of current limiting as the injector is held open.
5	This section of the waveform represents the main injection, which provides the majority of the combustible fuel used in the production of the power stroke. Depending on engine loads and demand, this section of the waveform will expand and contract as the engine speeds up and slows down.
6	This point on the waveform represents a small amount of current limiting as the injector is held open.
7	This period represents the total injection time for one cycle which includes three phases of injection: two pilot-injections and one main.

Automotive Actuators & Waveform Analysis

Piezoelectric Diesel injector

Unlike a solenoid Diesel injector, which is powered to open and closed with a return spring, a piezoelectric injector is powered both open and closed. A series of thin piezoelectric crystal wafers are stacked above the injector needle, and when provided with electric current, they deform and rapidly open the injector. Once open, the injector will remain in this state until a current with the opposite polarity is applied.

Care should be taken when diagnosing piezoelectric Diesel injectors. Because they are powered open and closed, it is possible that if an injector is disconnected from the circuit while the engine is running, it could be held in the open position, and excess fuel may lead to catastrophic failure.

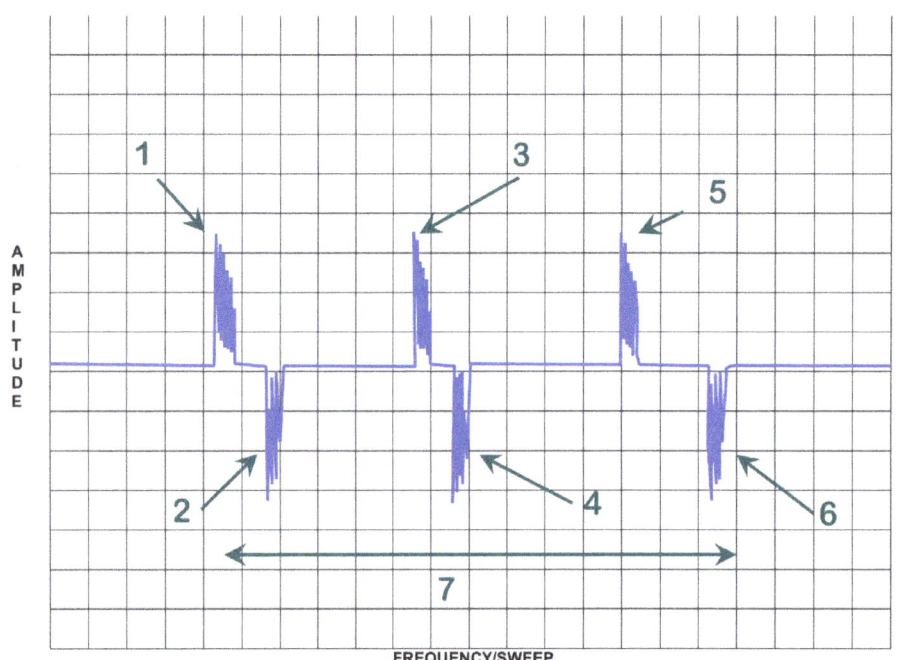
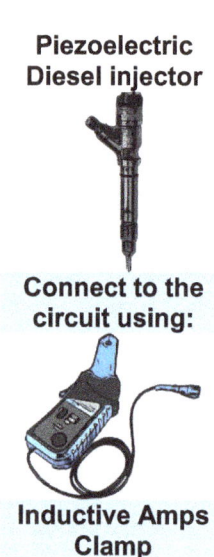

Figure 3.21 CRD piezoelectric injector current waveform

Automotive Actuators & Waveform Analysis

Table 3.21 Waveform analysis CRD piezoelectric injector current

Waveform component	Description
1	This is the point on the waveform where electric current is switched on, the piezoelectric wafers deform, opening the injector beginning the pilot phase of injection.
2	This is the point on the waveform where polarity is reversed, and the injector is closed, ending the pilot injection period.
3	This is the point on the waveform where the current is switched on for a second time, the piezoelectric wafers deform, opening the injector and the main phase of injection begins.
4	This is the point on the waveform where polarity is reversed and the injector is closed, ending the main injection phase. The duration of the main injection phase will vary depending on engine load and speed.
5	This is the point on the waveform where electric current is switched on for a third time, the piezoelectric wafers deform, opening the injector and the post phase of injection begins.
6	This is the point on the waveform where polarity is reversed and the injector is closed, ending the post injection phase.
7	The period shown is the complete injection duration to include, pilot, main and post injection phases.

Pumpe-Düse (PD)

PD or Pumpe-Düse are a style of electro-mechanical injector which develops its own high pressures inside the unit itself.

Unlike a common rail system, there is no high-pressure pump, but instead, the injector is acted on by a rocker arm from the camshaft; very high fuel pressures are created close to the point of injection. When injection is required, the operating solenoid is switched on by the positive circuit using battery voltage.

The following image shows three channels used to display the current and voltage waveforms from a Pumpe-Düse type Diesel injector.
- Channel 1 shows the injector current.
- Channel 2 shows the injector voltage supply circuit (the switching circuit).
- Channel 3 shows the injector earth circuit.

Using this method, you have the opportunity to compare the waveforms and check for any anomalies which may cause operating issues.

The injectors of a Pumpe-Düse system will be located inside the camshaft/rocker cover housing and will not be visible without some form of dismantling.
It is therefore important to have access to a good wiring diagram, to ensure that the correct wires are connected to the oscilloscope during testing.

Automotive Actuators & Waveform Analysis

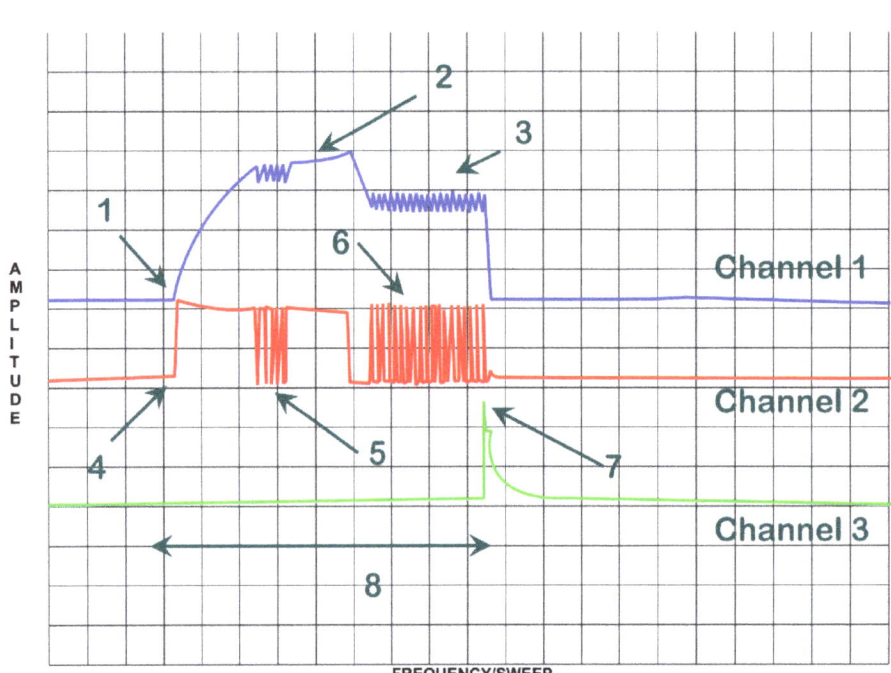
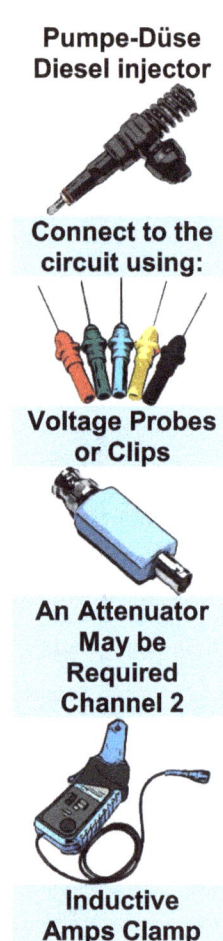

Figure 3.21 Pumpe-Düse injector current verses voltage

Table 3.21 Waveform analysis Pumpe-Düse Injector

Waveform component	Description
1	Channel 1. This is the point on the waveform where the injector is switched on and the current starts to flow, rising in a steady curve.
2	Channel 1. At this point, the current is being modulated to create the pilot injection phase.
3	Channel 1. At this point the current is being modulated to create the main injection phase.
4	Channel 2. This type of PD injector is switched through the live circuit, and shows the point at which the voltage is switched on, activating the injector.
5	Channel 2. At this point, the voltage is being modulated to create the pilot injection phase.

Automotive Actuators & Waveform Analysis

Table 3.21 Waveform analysis Pumpe-Düse Injector

6	Channel 2. At this point the voltage is being modulated to create the main injection (multi-pulsed) phase.
7	Channel 3. This shows the induced voltage in the earth circuit as the injector is switched off and the magnetic field in the injector windings collapses.
8	All channels. This is the total injection period to include both pilot and main injection.

Quantity control valve

The purpose of the quantity control valve in a common rail Diesel system is to regulate the amount of fuel flowing in the low-pressure circuit, between the feed pump in the tank and the high-pressure engine driven pump. Regulating this flow can help maintain pressure in the common rail between set limits.

The quantity control valve is often supplied with system (battery) voltage, and the earth circuit is then duty cycled to regulate the flow of Diesel.

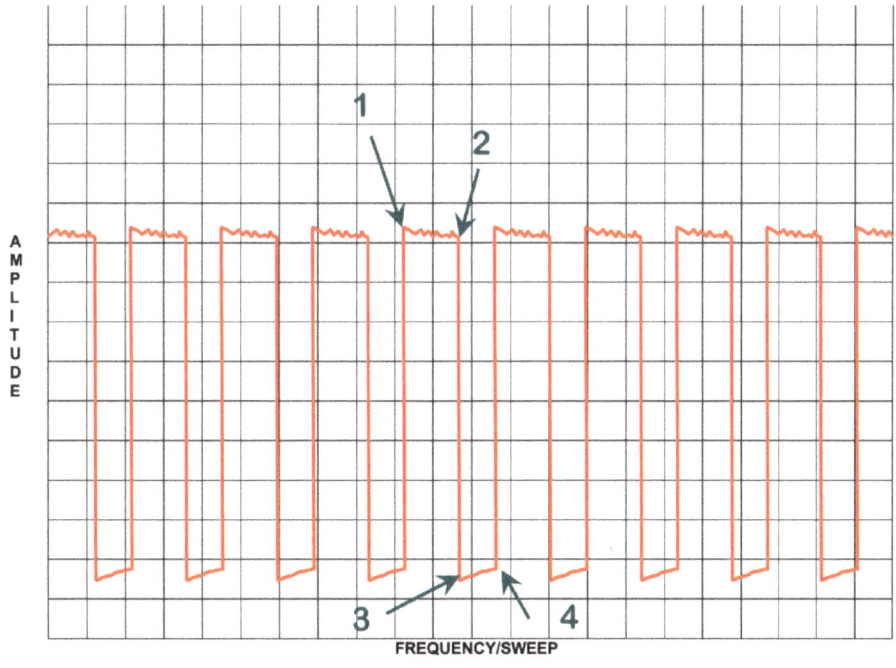

Figure 3.22 CRD quantity control valve voltage

Automotive Actuators & Waveform Analysis

Table 3.22 Waveform analysis quantity control valve (Diesel)

Waveform component	Description
1	This is the point on the waveform which shows the switched off period and is measured as a percentage of the total on/off period. It is not unusual to have small amounts of interference on the waveform at this point.
2	This is the point on the waveform where the quantity control valve is switched on and opens.
3	This is the point on the waveform which shows the switched-on period and is measured as a percentage of the total on/off period. It is not unusual to have small amounts of interference on the waveform at this point.
4	This is the point on the waveform where the quantity control valve is switched off and closes.

Throttle servo motor

On many drive-by-wire engines the mechanical throttle cable has been removed; a servo motor controls the throttle butterfly of a petrol engine. A two-channel waveform can be used to compare the voltage from the throttle position sensor and the duty cycle to the servo motor.

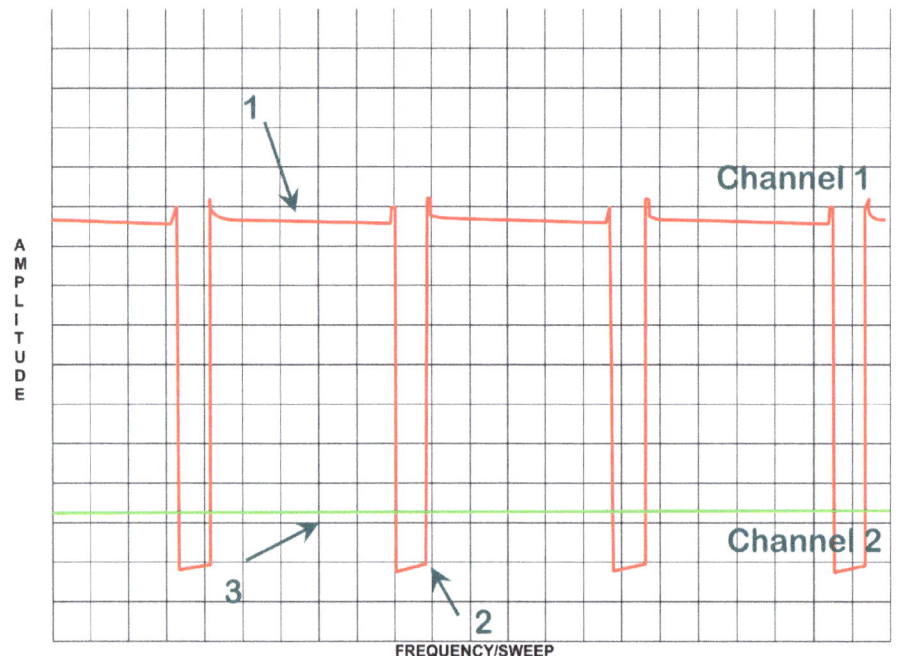

Figure 3.23 Throttle servo at idle

Automotive Actuators & Waveform Analysis

Table 3.23 Waveform analysis throttle servo (at idle)

Waveform component	Description
1	Channel 1. This is the point on the waveform which shows the switched-off period of the servo motor and is measured as a percentage of the total on/off period. The switched off period is relatively long, due to the fact the vehicle is at idle in this example.
2	Channel 1. This is the point on the waveform which shows the switched-on period of the servo motor and is measured as a percentage of the total on/off period. The switched-on period is relatively short, due to the fact the vehicle is at idle, with only partial butterfly opening in this example.
3	Channel 2. This shows a steady, relatively low voltage from the pedal position sensor, indicating that the vehicle is idling.

As engine revs are increased, the voltage from the throttle pedal position sensor on channel two will rise up the scale, and the duty cycle 'on period' will increase, opening the throttle butterfly wider.

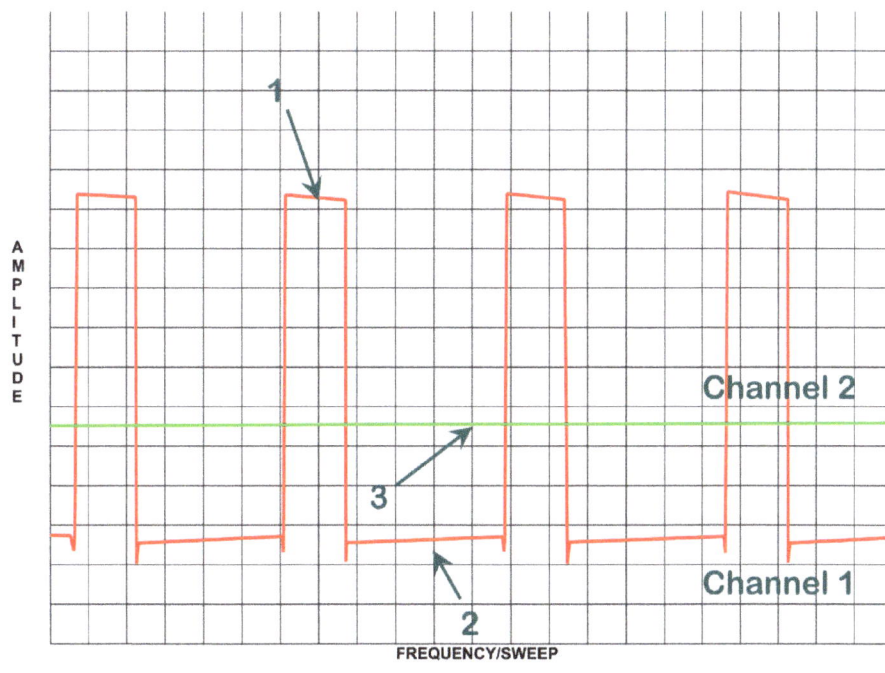

Figure 3.24 Throttle servo at 2,500 RPM

Automotive Actuators & Waveform Analysis

Table 3.24 Waveform analysis throttle servo (2,500 RPM)

Waveform component	Description
1	Channel 1. This is the point on the waveform which shows the switched-off period and is measured as a percentage of the total on/off period. The switched off period is relatively short, due to the fact the vehicle is revving in this example.
2	Channel 1. This is the point on the waveform which shows the switched-on period and is measured as a percentage of the total on/off period. The switched-on period is relatively long, due to the fact the vehicle is revving in this example.
3	Channel 2. This shows the voltage from the pedal position sensor has risen up the display, indicating that the vehicle is revving.

Cooling fan

Some manufacturers are producing variable speed electric cooling fans. This allows more flexibility in the design of engine cooling systems and better control over engine temperatures. The speed control of the cooling fan is achieved by duty cycle. (This example can be used as a basis for other vehicle cooling fans).

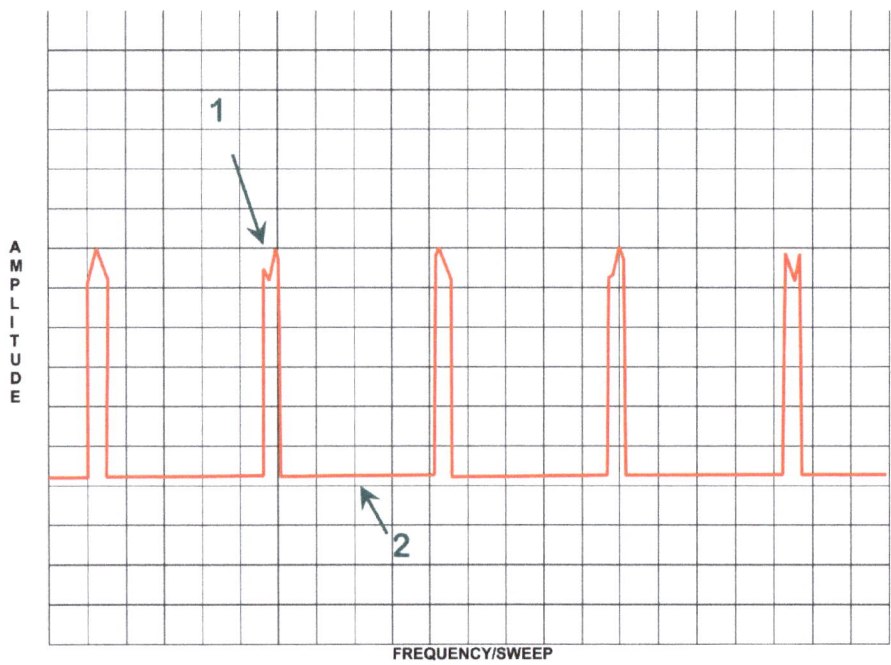

Figure 3.25 Electric cooling fan voltage

Automotive Actuators & Waveform Analysis

Table 3.25 Waveform analysis electric cooling fan

Waveform component	Description
1	This is the point on the waveform which shows the switched-off period and is measured as a percentage of the total on/off period. It is not unusual to have small amounts of interference on the waveform at this point.
2	This is the point on the waveform which shows the switched-on period and is measured as a percentage of the total on/off period.

VVT control

A large proportion of vehicle manufacturers are using some form of variable valve control in their engine design; this gives them flexibility of operation, while maintain good performance, emissions and fuel economy.

Many variable valve control systems used in engine design rely on oil pressure sent to the VVT actuator to operate the mechanisms, which alter valve phasing or lift. An ECU controlled solenoid valve (often supplied with battery voltage) can regulate the oil flow and pressure to the actuator mechanisms using duty cycle.

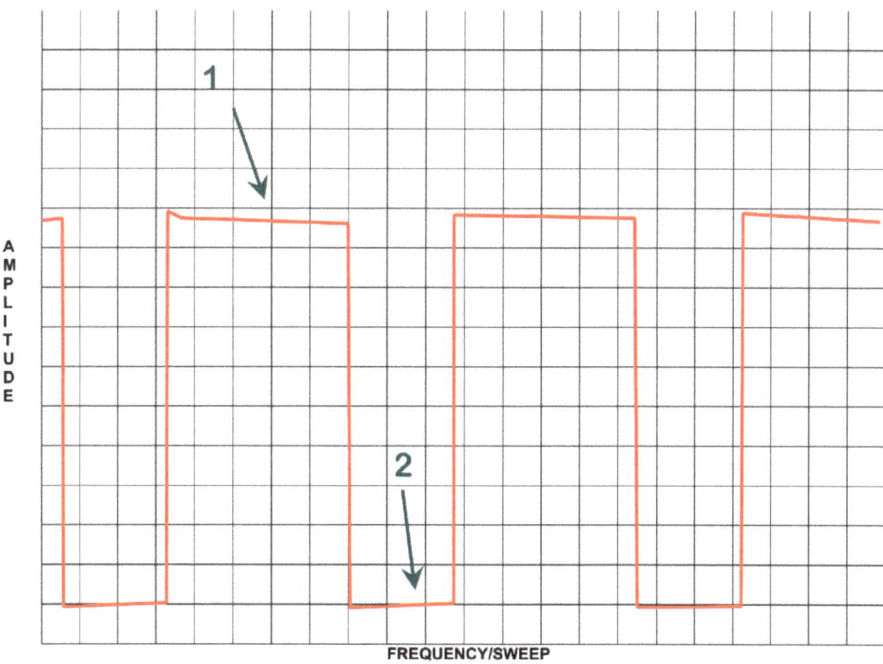
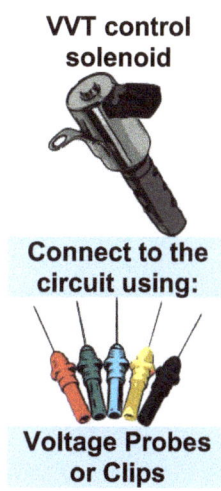

Figure 3.26 VVT control solenoid voltage

Automotive Actuators & Waveform Analysis

Table 3.26 Waveform analysis VVT control solenoid

Waveform component	Description
1	This is the point on the waveform which shows the switched-off period and is measured as a percentage of the total on/off period.
2	This is the point on the waveform which shows the switched-on period and is measured as a percentage of the total on/off period.

If the VVT design is more complex (for example, both inlet and exhaust valves or camshafts), it is not unusual to have multiple solenoids actuating the valve control mechanisms. The waveforms from different solenoids can be compared by using separate channels on the oscilloscope.

It is worthwhile bearing in mind that, as inlet and exhaust valve operation are separate factors in the cycle of an internal combustion engine, the waveforms produced on each may not have the same duty cycle period as one another, or may appear 'out of phase'.

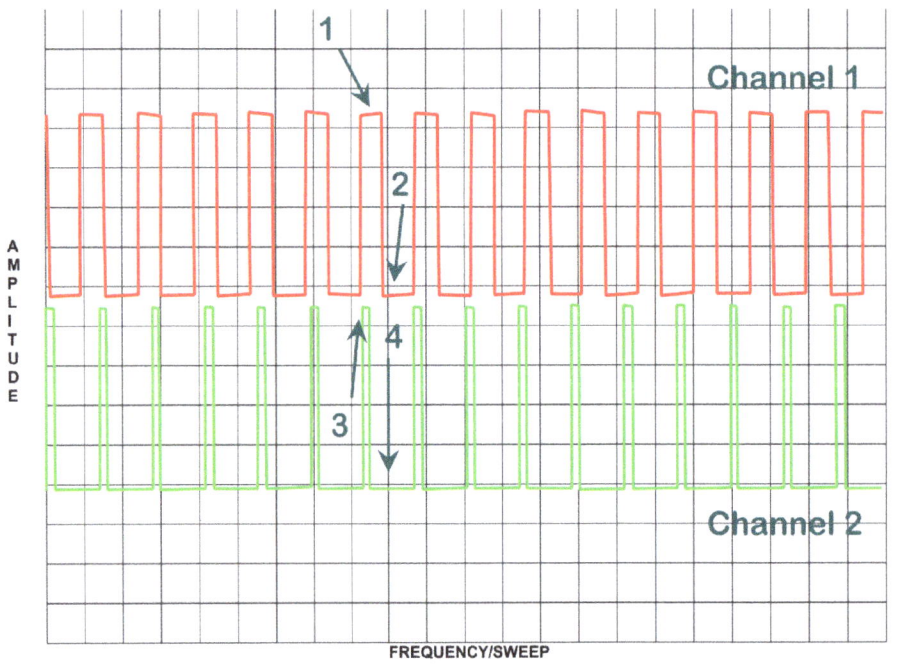

Figure 3.27 VVT double control solenoid voltage

Automotive Actuators & Waveform Analysis

Table 3.27 Waveform analysis VVT double control solenoid

Waveform component	Description
1	Channel 1. Inlet
This is the point on the waveform which shows the switched-off period and is measured as a percentage of the total on/off period.
2	Channel 1. Inlet
This is the point on the waveform which shows the switched-on period and is measured as a percentage of the total on/off period.
3	Channel 2. Exhaust
This is the point on the waveform which shows the switched-off period and is measured as a percentage of the total on/off period.
4	Channel 2. Exhaust
This is the point on the waveform which shows the switched-on period and is measured as a percentage of the total on/off period.

Starter motor

The starter is a powerful electric motor designed to rotate the crankshaft of the engine at speeds in excess of 180 RPM to initiate combustion.

A large amount of diagnostic information can be gathered about the motor and engine mechanical operation, if checked using an oscilloscope. By connecting a voltage probe and amps clamp the starter motor feed wire and cranking the engine, the waveforms produced can be analysed.

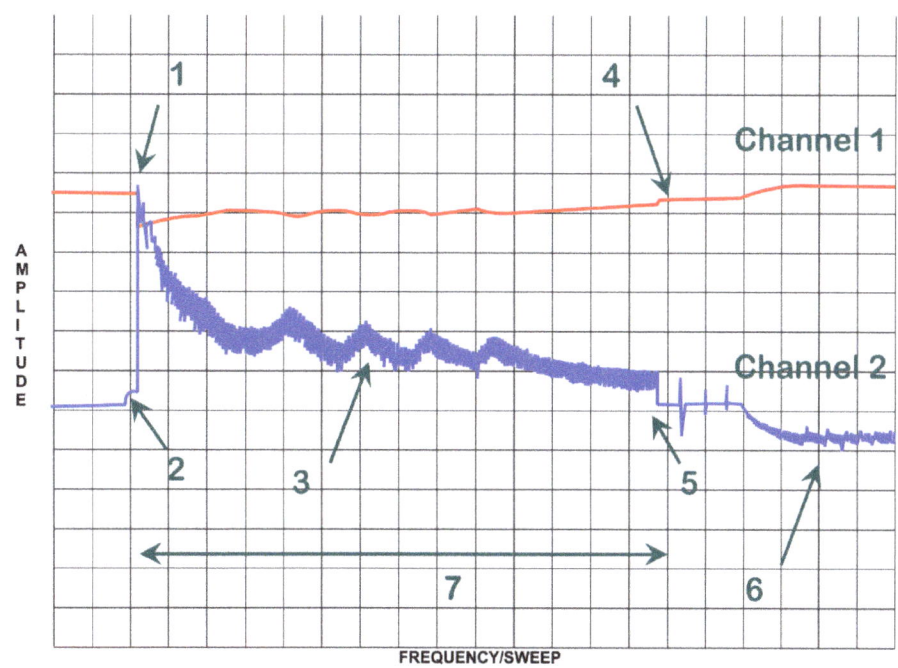
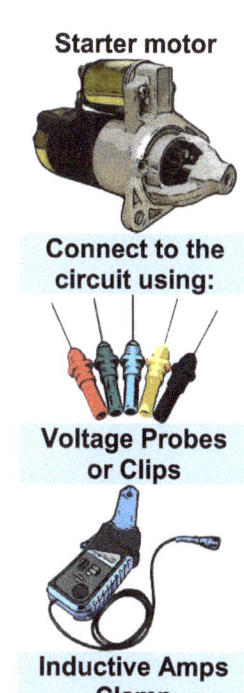

Figure 3.28 Starter motor current and voltage

Automotive Actuators & Waveform Analysis

Table 3.28 Waveform analysis starter motor current and voltage

Waveform component	Description
1	Channels 1 & 2. This is the start of cranking and can be seen in both the voltage waveform (channel 1) and current (channel 2). Voltage drops, caused by a potential difference on the system voltage as the starter motor initially cranks. Current rises rapidly at the start of cranking as the greatest effort placed on the starter motor happens as it tries to overcome the **inertia** taken to rotate the crankshaft. The initial peak current will often be far higher than the actual cranking current.
2	Channel 2. This small hump in the waveform indicates the current required to operate the starter solenoid when the key is turned or the starter button is pressed.
3	Channel 2. The rise and fall in the current at this point on the waveform is caused by the cranking compression pressures of the cylinders. The height of these peaks should be roughly similar if the relative compression of each cylinder is even.
4	Channel 1. At the end of the cranking period, voltage will initially return to battery values, before increasing slightly due to alternator charging.
5	Channel 2. This is the drop-off in current as the starter motor stops turning the crankshaft, when the ignition key or start button is released.
6	Channel 2. When the engine starts, it is possible for the starter motor to become a generator for a short period of time before it fully disengages from the flywheel ring gear. This can be seen in the reversed current dropping down at this point. Also, depending on the position of the current clamp, once the engine has started, current flow from the alternator may be shown as it recharges the battery.
7	Channels 1 & 2. This part of the waveform shows the total cranking period of the starter motor.

Inertia – the tendency for a body to resist acceleration or change of direction.

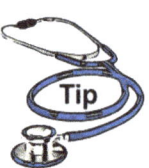

The current draw from a starter motor can be a good indication of the mechanical condition of the starter motor itself.
The amount of current needed will depend on the design of the motor and whether the engine is petrol or Diesel, however, a rough rule of thumb which can be used is that the initial cranking current should not exceed three times the amp/hour capacity of the battery.

Automotive Actuators & Waveform Analysis

 **Tip** It is not uncommon for a worn-out starter motor to create large amounts of electromagnetic interference (EMI) which can disrupt the operation of vehicle electronic control units (ECU).
If the waveforms from a starter motor show excessive 'noise' (interference), it may be advisable to recommend its replacement.

Relative compression

A current clamp placed around the starter motor feed wire can be used to provide an indication of engine relative compression. All peaks on the waveform should be relatively even:
- high peaks show good compression
- low peaks show poor compression

If a second channel is connected to the camshaft sensor, then by following the firing order, it will be possible to calculate which cylinder may be down on compression.

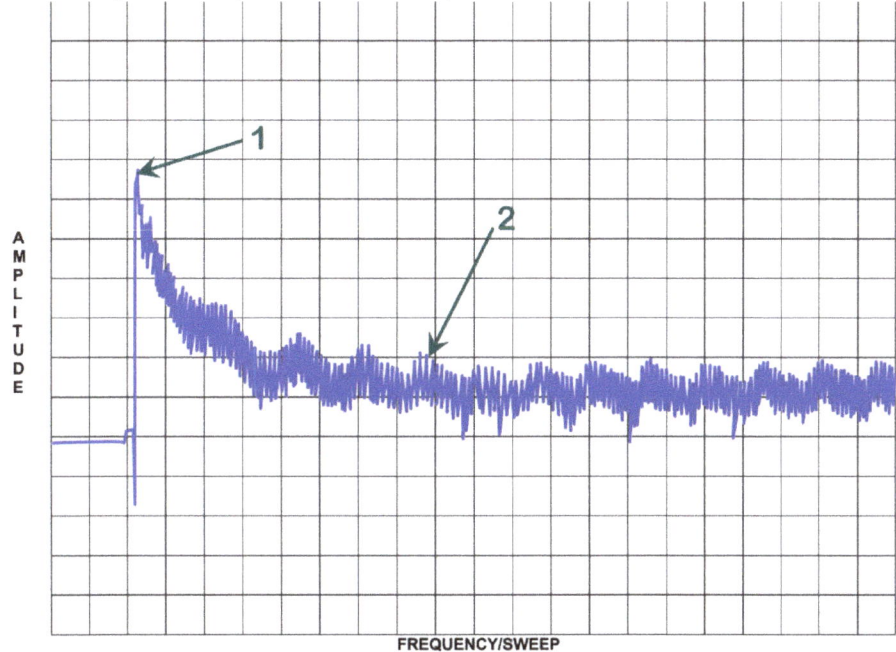

Figure 3.29 Relative compression

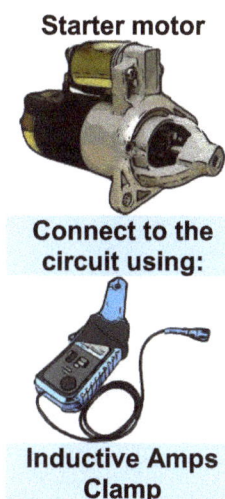

Starter motor

Connect to the circuit using:

Inductive Amps Clamp

Table 3.29 Waveform analysis relative compression

Waveform component	Description
1	Current rises rapidly at the start of cranking as the greatest effort placed on the starter motor happens as it tries to overcome the inertia taken to rotate the crankshaft. The initial peak current will often be far higher than the actual cranking current.
2	The smaller peaks, shown as an example at point 2, represent the strain placed on the starter motor relating to the compression stroke of each individual cylinder. The peaks should be relatively even; any low sections may indicate a lack of compression.

Automotive Actuators & Waveform Analysis

Tip — A poor reading shown on a relative compression diagnosis should be followed up by a physical compression or leakdown test to confirm your results.

Notes — Although relative compression can be a quick method to help assess the condition of an engine, far more information can be gained about the operation and health of individual cylinders by using a pressure transducer.

The image shown in figure 3.30 is of a starter motor on a Diesel engine and helps provide a comparison with one from a petrol engine.

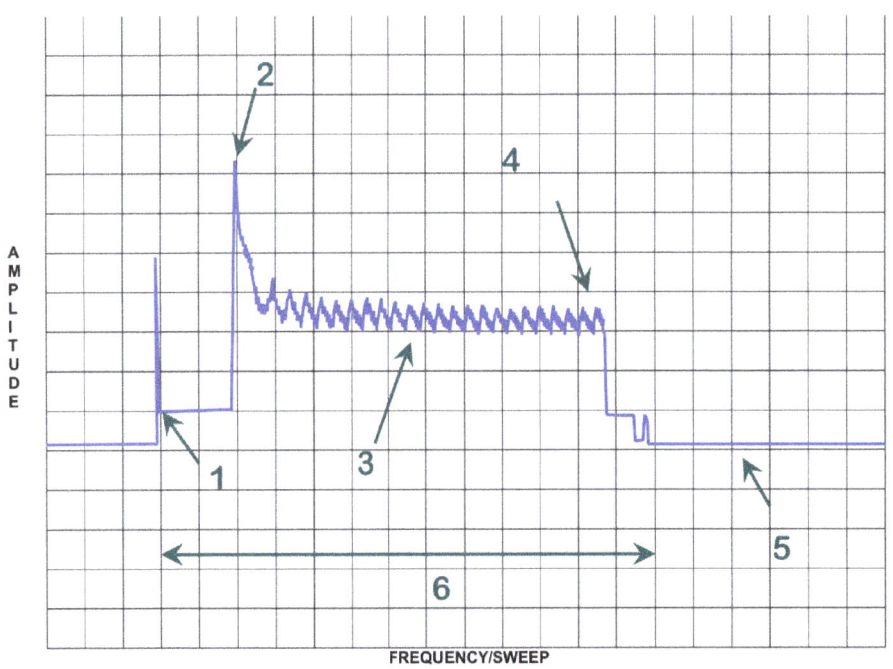
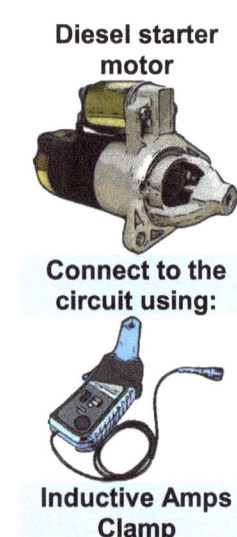

Figure 3.30 Diesel starter motor current

Automotive Actuators & Waveform Analysis

Table 3.30 Waveform analysis Diesel starter motor current

Waveform component	Description
1	This spike in the waveform indicates the current required to operate the starter solenoid when the ignition key or start button is pressed.
2	Current rises rapidly at the start of cranking as the greatest effort placed on the starter motor happens as it tries to overcome the inertia taken to rotate the crankshaft. The initial peak current will often be far higher than the actual cranking current.
3	The rise and fall in the current at this point on the waveform is caused by the cranking compression pressures of the cylinders. The height of these peaks should be roughly similar if the relative compression of each cylinder is even. This is a very useful diagnostic method, as conducting conventional compression testing on a Diesel engine can be awkward and time consuming.
4	This is the drop-off in current as the starter motor stops turning the crankshaft when the ignition key or start button is released.
5	At this point, following engine start, the waveform shows system operating current.
6	This part of the waveform shows the total cranking period of the starter motor.

Alternator

The alternator generates electrical energy to keep the battery charged, however, people often forget that it also provides the operating current for consumers when the engine is running.
Correct output is vital to ensure that all electrical systems operate as designed.

To check that the current is sufficient and that phases inside the alternator have not failed, it will be necessary to load the electrical system as current flow is produced by electrical demand.
When load is created by switching on electrical consumers such as headlamps and heated rear window systems, the amplitude of the waveform should increase.

It is possible to roughly calculate by how much the current draw should increase by using power law.

Wattage ÷ Voltage = Current

For example:
If two headlamp bulbs of 60 watts each are switched on, then 120 watts ÷ 12 volts = 10 amps' current.

Another method that can be used to place load on the alternator is to partially drain the battery. This way, when the engine is started the alternator should produce a high current output in order to try and top the battery back up as quickly as possible.

The current and voltage available during charging will vary from manufacturer to manufacturer and will also depend on the type of charging system fitted (see smart charging).

Automotive Actuators & Waveform Analysis

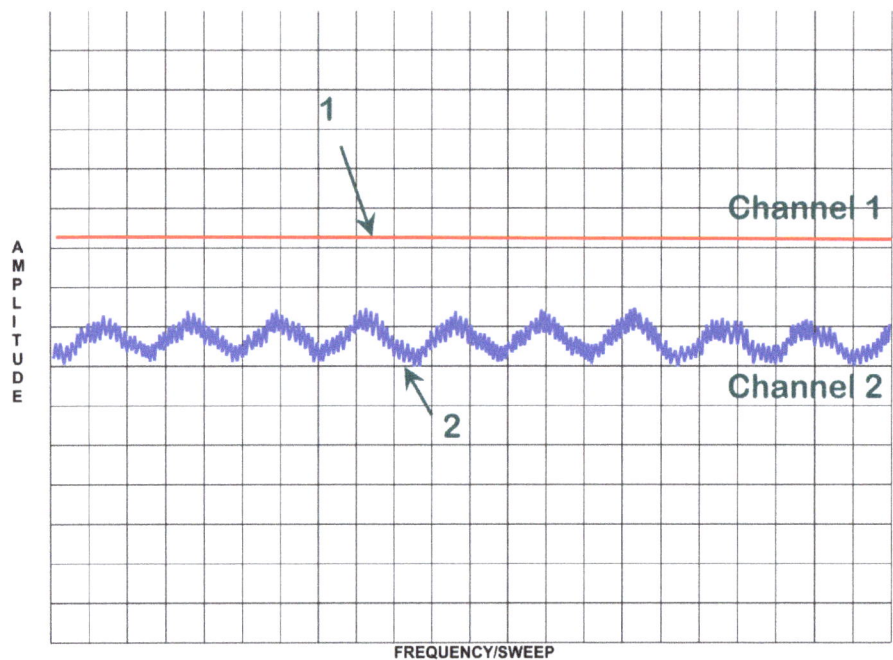
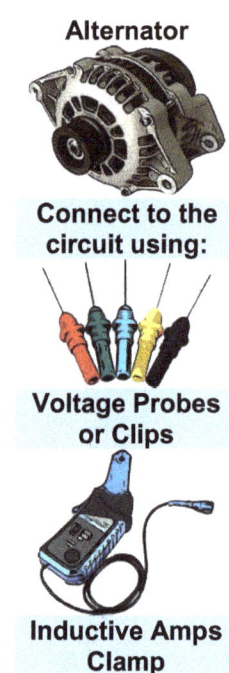

Figure 3.31 Alternator currant and voltage

Table 3.31 Waveform analysis alternator (current and voltage)

Waveform component	Description
1	Channel 1. This represents the regulated voltage, which can vary from manufacturer to manufacturer, but will typically be between 13.5 volts and 15 volts for a 12 volt system. (Its actual amplitude can be calculated using cursors on the oscilloscope screen if available).
2	Channel 2. This represents the current generated by the alternator and shows the characteristic ripple effect which is caused as the alternating current is rectified to direct current. The amplitude of this wave will vary depending on how much electrical load is placed on the system. (i.e. switching things on and off).

Tip: Switch on the headlamps without the engine running and leave for around ten minutes. This should allow time for the battery to partially discharge, and when the engine is started, the alternator should put out a higher current than normal by trying to recharge the battery as quickly as possible.

Automotive Actuators & Waveform Analysis

Rectified alternator ripple

It is also possible to check the rectified voltage output from an alternator, confirming that the windings in the stator and the diodes in the rectifier pack are functioning correctly.

To do this the oscilloscope should be connected across the battery positive and negative, however, unlike other tests, the voltage scale should be set to alternating current (AC). This way, when the engine is started and run, it will produce a waveform similar to that shown in figure 3.32.

You can see from the example, that the waveform isn't flat, but creates a distinctive ripple effect caused by the alternating current (AC) being rectified to direct current (DC).
The alternator has three internal windings formed in the stator at 120° intervals, so as the rotor turns through one complete revolution 360° it creates three distinct phases of charge. Should the diode rectifier or one of the stator windings fail, the output from the alternator will be reduced by 33%, but if checked with a multimeter this fault could be missed.

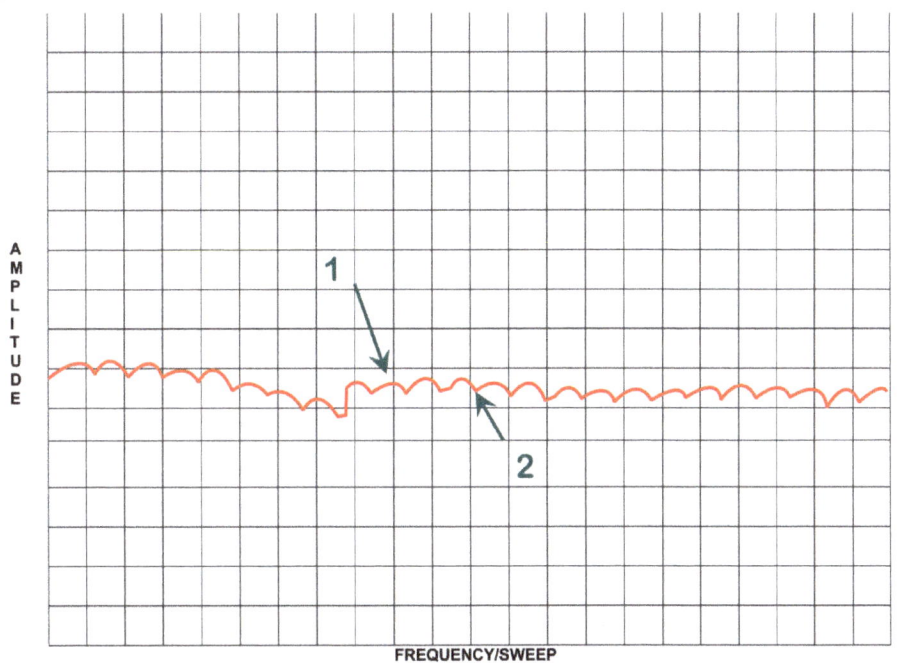

Figure 3.32 Alternator rectified voltage

Table 3.32 Waveform analysis alternator rectified voltage

Waveform component	Description
1	This shows the amplitude of the charging wave and is the rectified direct current DC voltage. The base to peak will often be less than 1 volt on a correctly operating alternator. If the alternator is faulty, the base to peak voltage will be in the range of 3 to 4 volts.
2	This shows short base spikes as the voltage is rectified to direct current DC and should not drop down into a negative value. If the alternator is faulty, long downwards tails would appear every third wave at regular intervals, showing that one phase is not being correctly rectified.

Automotive Actuators & Waveform Analysis

Smart charge alternator

It has long been known that if a vehicle battery can be maintained at above 80% electrical capacity, lifespan and loads on electrical systems can be considerably reduced, leading to better performance, fuel economy and lower emissions.
Some manufacturers are now producing vehicles with 'smart charge' systems, where an alternator control unit can estimate the state of charge (SOC) of the battery and regulate the output to maintain a relatively stable voltage and current depending on electrical demands.

In order to test a smart charge alternator, it will be necessary to determine the function of each wire at the alternator connection. This could be done using a wiring diagram, or worked out by analysing the patterns produced on the oscilloscope.

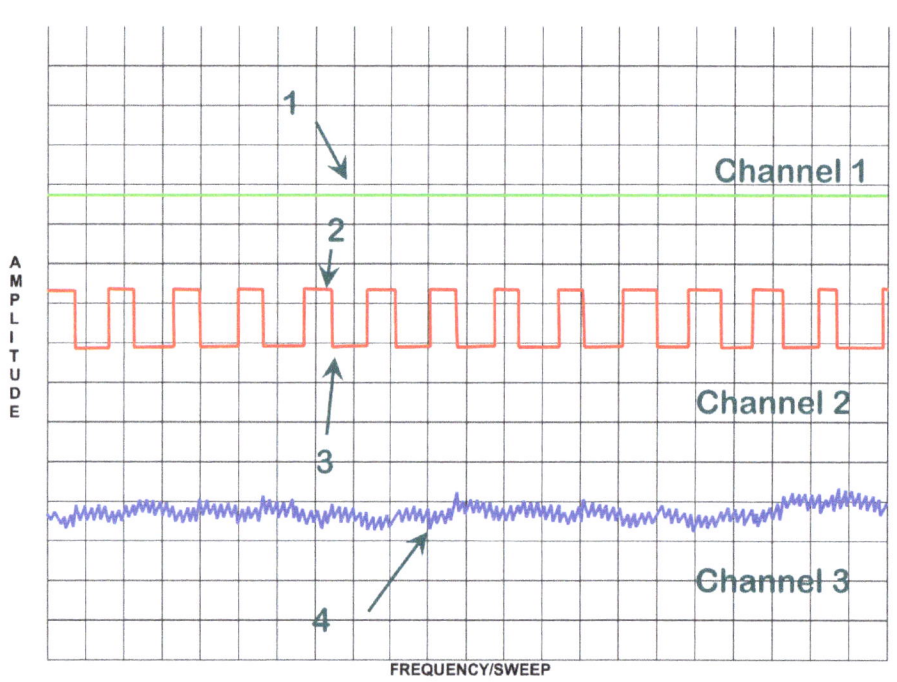
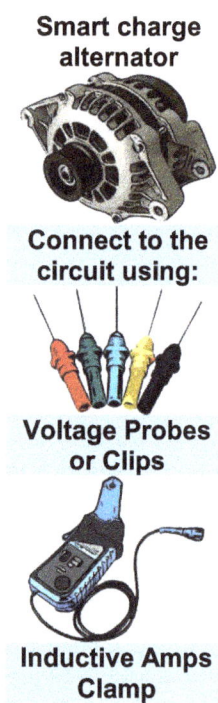

Figure 3.33 Smart charge alternator waveforms (no load)

Table 3.33 Waveform analysis smart charge alternator (no load)

Waveform component	Description
1	Channel 1. When a voltage probe is connected to the load command (request) wire from the ECU to the alternator, this will produce a duty cycle that regulates the voltage being generated in the alternator. With no electrical loads placed on the alternator, this waveform is idle and producing no command request; however, this should change if electrical demands are placed on the charging system.

Automotive Actuators & Waveform Analysis

Table 3.33 Waveform analysis smart charge alternator (no load)

2	Channel 2. When a voltage probe is connected at the feedback pin to the ECU from the smart charge alternator, a square waveform duty cycle should be seen helping the ECU determine the regulated voltage. This section shows the off period of the duty cycle and will vary depending on the output of the regulated voltage from the alternator.
3	Channel 2. When a voltage probe is connected at the feedback pin to the ECU from the smart charge alternator, a square waveform duty cycle should be seen helping the ECU determine the regulated voltage. This section shows the on period of the duty cycle and will vary depending on the output of the regulated voltage from the alternator.
4	Channel 3. When a current clamp is placed on the output wire of the alternator (<u>not the battery wire</u> as this would display the balance of current being produced by the alternator and the current consumed by the vehicle's electrical systems) it will show the current being generated by the alternator. When there is no load on the system, the amplitude will be relatively low.

> Ford smart charge alternators usually have a three-pin plug connecting them with the charging ECU.
> Following the manufacturers wiring diagram the pins will include:
> 1 * Alternator Feedback
> 2 * Alternator Load Request
> 3 * Reference Voltage

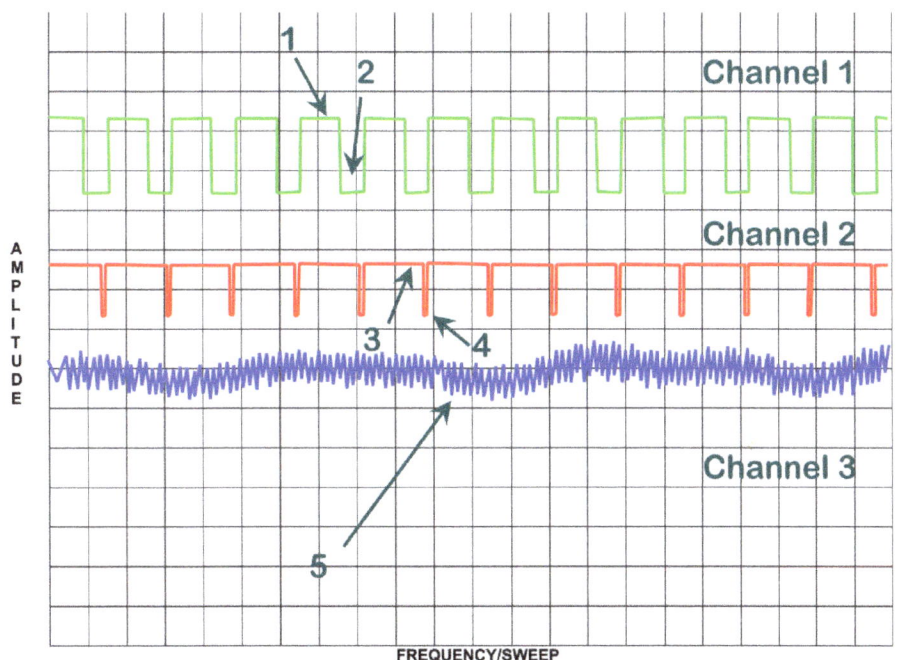

Figure 3.34 Smart charge alternator waveforms (under load)

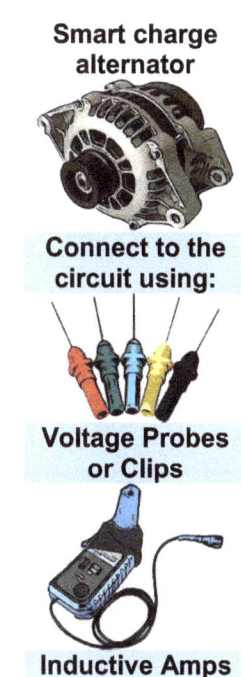

Automotive Actuators & Waveform Analysis

Table 3.34 Waveform analysis smart charge alternator (loaded)

Waveform component	Description
1	Channel 1. When a voltage probe is connected to the load command (request) wire from the ECU to the alternator, this will produce a duty cycle that regulates the voltage being generated in the alternator. With electrical loads placed on the alternator, this waveform is producing a command request duty cycle, however, this should change if electrical demands are taken away from or added to the charging system. This section shows the off period of the duty cycle.
2	Channel 1. When a voltage probe is connected to the load command (request) wire from the ECU to the alternator, this will produce a duty cycle that regulates the voltage being generated in the alternator. With electrical loads placed on the alternator, this waveform is producing a command request duty cycle, however, this should change if electrical demands are taken away from or added to the charging system. This section shows the on period of the duty cycle.
3	Channel 2. When a voltage probe is connected at the feedback pin to the ECU from the smart charge alternator, a square waveform duty cycle should be seen helping the ECU determine the regulated voltage. This section shows the off period of the duty cycle and will vary depending on the output of the regulated voltage from the alternator.
4	Channel 2. When a voltage probe is connected at the feedback pin to the ECU from the smart charge alternator, a square waveform duty cycle should be seen helping the ECU determine the regulated voltage. This section shows the on period of the duty cycle and will vary depending on the output of the regulated voltage from the alternator.
5	Channel 3. When a current clamp is placed on the output wire of the alternator (<u>not the battery wire</u> as this would display the balance of current being produced by the alternator and the current consumed by the vehicle's electrical systems) it will show the current being generated by the alternator. When there is load on the system, the amplitude will be relatively high.

Ford smart charge alternators usually have a three-pin plug connecting them with the charging ECU.
If the alternator does not appear to be charging, the plug can be disconnected and the engine started (<u>do not disconnect while the engine is running</u>) and the alternator will revert to a standard charging system. This may help you isolate different charging issues, but should not be considered a long-term solution to the fault.

Automotive Actuators & Waveform Analysis

Never jump start vehicles that use smart charging systems, because if the charge control module experiences a battery in very poor condition, and a cold engine, it may initially provide a very high voltage in order to try and recover the battery. High voltages of around 18 volts, may be enough to cause system damage to vehicle electronics and ECU's.

Smart charge systems do not work with standard lead acid batteries.
It is important that if a smart charge system is used, the vehicle battery is a silver calcium type to ensure correct operation.

ABS solenoid

The ABS solenoid controls a hydraulic valve which helps regulate fluid pressure in the braking system. Depending on how much current is supplied to the solenoid, the valve can be positioned to allow an increase in pressure, a holding pressure or a reduction in pressure. These three stages of operation form the basis of an anti-lock braking system which is trying to prevent the wheel from skidding in an emergency situation. The position of the solenoid control valve is governed by rapidly pulsing the voltage on and off, therefore regulating the current that is flowing in the solenoid winding and the strength of the magnetic field produced. To do this the ECU will use duty cycle or pulse width modulation (PWM).

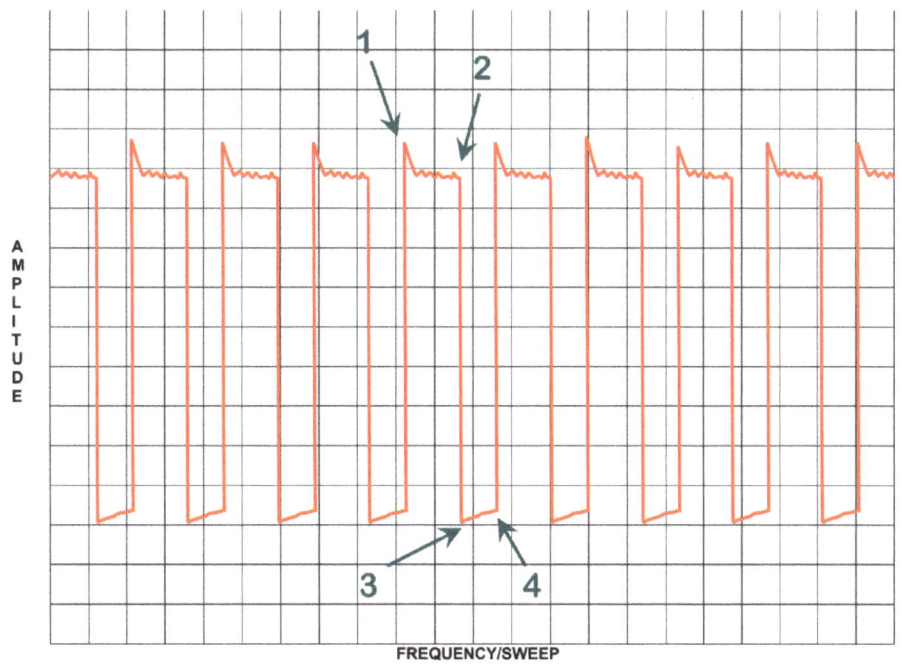

Figure 3.35 ABS solenoid waveform

Automotive Actuators & Waveform Analysis

Table 3.35 Waveform analysis ABS solenoid

Waveform component	Description
1	This is the point on the waveform where voltage has been switched-off. Due to the collapse of the magnetic field generated inside the solenoid winding coil, it is not uncommon to see an induced voltage spike at this point.
2	This is the point on the waveform where the solenoid is switched-on and voltage rapidly falls.
3	At this point on the waveform, voltage will be initially at around 0 volts as the solenoid has been switched to ground, but you may see a small amount of voltage rise away from zero as current starts to flow.
4	This is the point on the waveform where the solenoid is switched off and voltage once again returns to battery (system) voltage.

If the ABS system uses pulse width modulation to control, the solenoids, the waveform will be similar in shape to a petrol fuel injector pattern.

Automotive Sensors
& Waveform Analysis

Chapter 4 Automotive Sensors and Waveform Analysis

This chapter will help you develop knowledge and understanding of automotive sensors. It will enable you to conduct effective diagnosis and repairs of system faults; supporting you by providing a breakdown of waveform images and analysing why the patterns are formed. Remember to work in a systematic way, and observe the relevant environmental, health and safety regulations at all times.

Contents

Introduction	85
Sensors and their purpose	86
Wheel speed sensors	93
Throttle pedal sensors	98
Vane air flow meter	99
Mass air flow meter	101
Diesel air flow meter	102
Vortex air flow meter	103
Manifold absolute pressure sensor and intake pressure sensor	104
Crankshaft sensor	108
Coolant temperature sensor	110
Air temperature sensor	112
Barometric pressure sensor	113
Camshaft sensors	114
Fuel pressure sensor	117
Knock sensor	118
Lambda sensor + heater	120
Vehicle speed sensor VSS	127
Sun load sensors	128
Steering angle and torque sensor	129

Automotive Sensors & Waveform Analysis

There are many hazards associated with the service and maintenance of light vehicle electrical and electronic systems. You should always assess the risks involved with any diagnostic, maintenance or repair routine before you begin and put safety measures in place.
You need to give special consideration to the possibility of:
• The risk of electric shock.
• The hazards associated with running engines in confined spaces.
You should always use appropriate personal protective equipment (PPE) when you work on these systems. Make sure that your selection of PPE will help protect you from these hazards.

Don't forget your PPE and VPE

  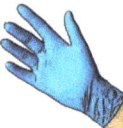 

 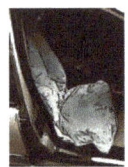  

Information sources

The complex nature of light vehicle electric and electronic systems requires a good source of technical information and data. In order to conduct diagnostic, maintenance and repair procedures, you need to gather as much information as possible before you start.

Sources of information may include:

Table 4.1 Possible information sources

Verbal information from the driver	Vehicle identification numbers
Service and repair history	Warranty information
Vehicle handbook	Technical data manuals
Workshop manuals/Wiring diagrams	Safety recall sheets
Manufacturer specific information	Information bulletins
Technical helplines	Advice from other technicians/colleagues
Internet	Parts suppliers/catalogues
Jobcards	Diagnostic trouble codes
Oscilloscope waveforms	On vehicle warning labels/stickers
On vehicle displays	Reference/Textbooks

Always compare the results of any inspection, testing or diagnosis to suitable sources of data. Remember that no matter which information or data source you use, it is important to evaluate how useful and reliable it will be to your safety, diagnostic, maintenance and repair routine.

Automotive Sensors & Waveform Analysis

Where to start?

Step 1
- A good systematic diagnostic routine should always begin with careful questioning of the driver to gather as much information about the symptoms and history of the fault as possible.

Step 2
- After a brief visual inspection for obvious signs of damage or safety issues, the system should be tested to try and recreate the fault.

Step 3
- The vehicle should then be scanned for diagnostic trouble codes and any codes should be recorded. (If possible, a full scan should be conducted, as issues in unrelated systems can sometimes affect the operation of others).

Step 4
- Any codes should be cleared and the vehicle should be tested over a complete drive cycle.

Step 5
- Rescan the vehicle and concentrate diagnosis around any codes that have returned.

Step 6
- Connect the oscilloscope to the suspected circuit and analyse any waveforms produced.

Introduction

Sensors provide input to the ECU, helping it monitor vehicle systems. They have different functions depending on their location and the type of data to be gathered, but will either be active or passive, inductive or voltage fed.

It is worthwhile remembering that a systematic diagnostic routine can help reduce wasted time, excessive effort and frustration. As a result, it is always good practice to test sensor inputs at two places, the sensor and the ECU harness plug. Checking signals at the sensing components will tell you if signals are being generated correctly. Checking at the harness plug will tell you if the signals are getting through to their destination, or if there is any miscommunication on the way. Careful analysis of the waveform should always be conducted, as a great deal of information can be obtained which can help with a successful diagnosis, leading to a first-time fix.

An important point to remember when testing sensors is that some of them may generate their own voltage, (inductive) while others will be powered by the system ECU. It is important to understand how the sensor functions in order to distinguish any differences and form a satisfactory conclusion on its operation.

Automotive Sensors & Waveform Analysis

Switches

Some electronic systems use switches to signal a physical change in a system or component; this is a binary form of sensing.
A switch is simply a method of making or breaking an electric circuit.
Unlike other forms of sensors which can operate through a range of electrical control, a switch will be either on or off, and if measured using an oscilloscope, will produce a simple square wave when switched.

Active and passive sensors

There is often confusion surrounding the key terms active and passive sensors, and therefore this terminology should be used with caution when describing the function and operation for vehicle sensors.

A passive sensor is one that reacts (detects and responds) to some type of input from the physical environment including:
- Heat
- Light
- Radiation
- Magnetism
- Vibration
- Or pressure

An active sensor one that will often send out, or transmit a signal (using light or electrons for example) and bounce them off a target, gathering the reflected information which can then be converted into data.

These key terms should not be confused with sensors that make their own voltage (inductive) or those that are fed with power to work.

Inductive

An inductive sensor is one that uses a magnetic field to induce a voltage in an internal winding. These sensors create their own voltage when the magnetic field surrounding them is disrupted. Because they don't need to be provided with a power source, it is possible to test an inductive sensor while it is unplugged from the circuit, as long as a motion that causes its magnetic field to move can be simulated.

Voltage fed

A voltage fed sensor needs power from the vehicle's electrical system in order to work.
These sensors are often passive in nature and transmit a signal and receive reflected information.
It is very important when testing a voltage fed sensor with an oscilloscope to ensure that it is plugged in and the circuit is switched on, for it to give a waveform.

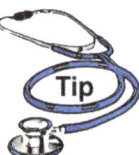

Both inductive and voltage fed sensors can be tested **in situ** with the circuit connected and switched on. It is therefore good diagnostic practice, where possible, to do this with all sensors to avoid misleading results.

Automotive Sensors & Waveform Analysis

In situ – still fitted to the car (not removed).

Examples of sensor type and a brief description of their purpose is shown in table 4.2.

Table 4.2 Examples of sensors and their purpose

Sensor	Description
Wheel speed sensors (inductive, MRE, Hall Effect) 	A wheel speed sensor is often mounted on or near the hub and is designed to indicate rotational speed. The signal produced can be used by a number of systems on the car, such as, **ABS**, **TRC**, **EBD**, **VSC** and many more. There are three main styles of wheel speed sensor in common use: • Inductive • **MRE** • Hall Effect
Throttle position sensors 	The throttle position sensor is mounted on the throttle body and is often a variable resistor, known as a **potentiometer**. When used in conjunction with a measurement of air flow or manifold pressure, this sensor can help give an indication of engine load. Because many cars are drive by wire (i.e. no throttle cable) most throttle position sensors have multiple potentiometer tracks as a back-up safety system, should one potentiometer fail. Some throttle position sensors incorporate a switch in their design which can signal the ECU when the throttle butterfly is at idle or wide open throttle (WOT).
Throttle pedal sensors 	As most vehicles are now **drive-by-wire** (i.e. no throttle cable, but instead controlled by motors and electronics) the position of the accelerator is vital for the ECU to calculate driver demand. For safety reasons, most accelerator pedal position sensors are multi-track potentiometers, providing back-up signals that can be compared by the ECU to ensure the reliability of the information provided. As the throttle pedal is pressed by the driver, an analogue signal is produced, showing a rising and falling voltage which is proportional to the position of the pedal and will react in accordance with driver inputs.
Vane air flow meter 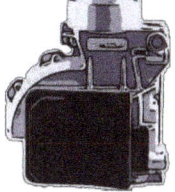	The vane type air flow sensor, is an older style air measurement device. Mounted in the induction system after the air filter, a flap connected to a potentiometer is moved by the air passing through it, and the corresponding resistance indicated by the potentiometer is proportional to the flow of air. Due to the restriction of air flow caused by the flap in these sensors, they have now been superseded by other types.

Automotive Sensors & Waveform Analysis

Table 4.2 Examples of sensors and their purpose

Sensor	Description
Mass air flow meter 	A mass air flow sensor (MAF) is a very accurate method of measuring the quantity of air entering the intake system and provides little resistance to the flow of air. A thin wire or electrically conductive film is heated using current. As air passes over the 'hot wire' it is cooled. The ECU must now supply more current in order to keep this wire at a constant temperature. The rise in current supply is proportional to the mass of air flowing through the sensor.
Vortex air flow meter 	A vortex air flow sensor is similar in size and shape to a mass air flow sensor. At the inlet to the sensor, a triangular shaped column is mounted vertically in the airstream. As air passes over this column it is split into two swirling whirlpools or vortices that turn in opposite directions. In the vortex sensing chamber, the disruption to the airflow caused by the swirling air currents is detected by a pressure or sonic pick-up and converted into a digital signal.
Manifold absolute pressure sensor (analogue and digital) 	A manifold absolute pressure sensor or MAP is another type of air measurement sensor. It is connected directly, or indirectly via a pipe, to the inlet manifold or plenum chamber. Its job is to measure the **depression** (low pressure) or vacuum found in the manifold during engine operation. If the depression is measured and the volume of the manifold is known, a calculation can be made which results in data showing the amount of air being inducted into the engine. MAP sensors can produce either analogue or digital signals, depending on their design.
Crankshaft sensor (inductive, Hall Effect, AC excited) 	The purpose of a crankshaft sensor is to indicate engine speed and crankshaft position. Mounted next to the crank pulley or flywheel, the sensor receives a pulsating signal from a form of toothed wheel which rotates past it. At strategic points in the toothed wheel, teeth are missing to give an indication of crankshaft position (normally top dead centre TDC). There are three main types of crankshaft sensor, inductive, Hall effect and AC excited. The inductive sensor requires no external power source to work, however, the Hall effect and AC exited do.
Coolant temperature sensor 	Coolant temperature sensors are a form of thermal resistor known as a **'thermistor'**. They will be mounted in the cylinder head or engine block where part of the sensor is exposed to the coolant. As the temperature of the coolant rises, there is a resistance change in the sensor, providing a variable voltage signal to the ECU which corresponds to the engine temperature. There are two main types **Negative Temperature Coefficient NTC** (the most common) and **Positive Temperature Coefficient PTC**. • With an NTC thermistor, as temperature rises, resistance falls. • With a PTC thermistor, as temperature rises, so does resistance.

Automotive Sensors & Waveform Analysis

Table 4.2 Examples of sensors and their purpose

Sensor	Description
Air temperature sensor 	An air temperature sensor helps the engine management system determine the density of the air entering the combustion chamber so that it can allow for this in its calculation of air/fuel ratio. (Cold air is denser than warm air and therefore contains more oxygen). The sensor may measure ambient temperature, the outside air temp, or it may measure the temperature of the air moving through a section of intake tract. If it is an intake air temperature measurement, it may be higher than ambient depending on its position in the circuit (after a turbo charger for example). There are two main types Negative Temperature Coefficient NTC and Positive Temperature Coefficient PTC. • With an NTC thermistor, as temperature rises, resistance falls. • With a PTC thermistor, as temperature rises, so does resistance.
Barometric pressure sensor 	A barometric (or Baro) sensor is normally a form of pressure transducer which helps the engine management system determine the density of the air entering the combustion chamber, so that it can allow for this in its calculation of air/fuel ratio. (High pressure air is denser and therefore contains more oxygen). Using a barometric pressure sensor will help the ECU determine altitude, which will also affect the air/fuel ratio.
Camshaft sensor (inductive and Hall Effect) 	The camshaft sensor is used in conjunction with the crankshaft sensor to provide the engine management system with an indication of which cylinder is at top dead centre TDC at any one time. Usually mounted near the camshaft pulley(s), it will pick up on a single tooth of a **reluctor** or pulse wheel every time number one piston is at TDC. Once the position of a single cylinder is known, by comparing it to the position of the crankshaft at the time, it can work out the effective position of the other pistons (i.e. cylinder recognition). Once the position of all of the cylinders is known, the engine management is now able to effectively control the ignition timing and fuel injection. Two main types of camshaft sensor are in common use: • Inductive • Hall effect
Distributor (inductive and Hall effect) 	Early ignition systems used a distributor to control the timing of the high voltage spark at the plugs. When conventional ignition systems replaced contact breakers with **transistors**, information was needed to indicate when the spark was required; the two main methods involved pick-ups located inside the distributor. An inductive pick-up works in a similar manner to a crankshaft position sensor, using a toothed reluctor to indicate when a spark is needed. (The reluctor has the same number of teeth as there are cylinders). A Hall effect sensor uses a slotted drum, or windows and doors, to interrupt the signal produced by the sensor. (The rotating drum has the same number of windows and doors as there are cylinders). An inductive pick-up will produce a sinewave, and a Hall effect will produce a square wave.

Automotive Sensors & Waveform Analysis

Table 4.2 Examples of sensors and their purpose

Sensor	Description
Fuel pressure sensor 	Used on both petrol and Diesel engines, the fuel pressure sensor is a form of pressure transducer. It provides a signal to the engine management unit which will help regulate injection time and maintain the correct air/fuel ratio. If fuel pressure is too high, the engine will run rich, if the fuel pressure is too low, the engine will run week.
Knock sensor 	Pinking or engine 'ping' is a condition where ignition timing is too far advanced. It occurs because, as the flame spreads following ignition, the pressure build-up on top of the piston crown happens when the firing cylinder is at top dead centre TDC. With the connecting rod completely vertical, energy is lost and a distinctive knock can be heard from the cylinder. (To make best use of the pressure created by the combustion process, the piston should be just after TDC). A knock sensor is mounted on the cylinder head or block, usually between two cylinders, and monitors the engine for pinking. If engine 'ping' occurs, the ignition timing is retarded in stages of two degrees until pinking stops. When engine 'ping' is no longer detected, the ignition timing is advanced in stages of one degree until pinking once again reoccurs. This way, ignition timing is always kept in the optimal operating window.
Lambda sensor + heater (Zirconia and Titania) 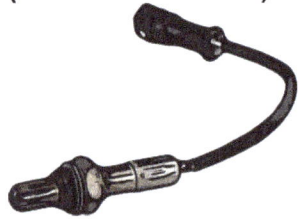	An exhaust gas oxygen sensor (sometimes known as a Lambda sensor) is mounted in the exhaust system before the catalytic converter. Its job is to monitor the oxygen content of the exhaust gas and help the engine management system check if it is operating within the **tolerance** of air/fuel ratio. If the engine is running weak, there will be too much oxygen left over after combustion. If the engine is running rich, there will not be enough oxygen left over after combustion. There are two main types of standard exhaust gas oxygen sensor, Zirconia and Titania. When these materials come into contact with the oxygen in exhaust gasses, they change their resistance and a corresponding signal is sent to the ECU. Many exhaust gas oxygen sensors also contain an electrically operated heating element to ensure that they reach their required operating temperature as soon as possible, and therefore work effectively. Wideband or broadband oxygen sensors are used for the same purpose as a standard lambda sensor, however, they use a component called a nerst cell to measure exhaust gas oxygen content. These sensors provide a much more stable signal that is proportional to oxygen content and are therefore more accurate. Post-catalyst oxygen sensors are used to determine the operating efficiency of the catalytic converter by comparing its readings with those of the pre-cat sensor. If the catalyst is operating efficiently, a greatly reduced signal will be measured post-cat as much of the oxygen left over from combustion has been used up in the catalytic reaction to reduce the harmful exhaust pollutants.

Automotive Sensors & Waveform Analysis

Table 4.2 Examples of sensors and their purpose

Sensor	Description
NOx sensor 	On vehicles with direct petrol fuel injection which operate in a fuel **stratified** 'lean burn' mode for economy and reduced emissions, the post-catalyst oxygen sensor is often replaced with an NOx sensor. This is because lean burn conditions create high combustion temperatures leading to the increased production of NOx (oxides of nitrogen). Visually an NOx sensor is very similar in design shape and size as a Lambda sensor, but it will have the ability to monitor the NOx content of the exhaust gas. A three-way catalytic converter contains a reduction **catalyst** which retains nitrogen to reduce the quantity of NOx released to atmosphere. If values of NOx rise beyond the pre-set tolerance, the sensor signals that the catalyst requires regeneration.
Vehicle speed sensor 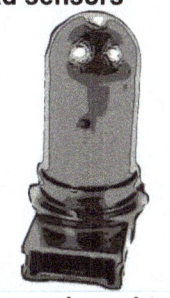	With modern **multiplexed** network systems, vehicle speed can be determined from several different sources, such as ABS wheel speed sensors; however, it's not uncommon for a dedicated component to be used for this purpose. Often mounted on the transmission output, they can be inductive or Hall effect with operation similar to a crankshaft sensor. Some vehicle speed sensors (VSS) have their own gear drive connected directly to the transmission and signalling is internal to the sensor itself.
Sun load sensors 	The sun load sensor is a small **photoelectric** cell, whose resistance changes in relation to the amount of sunlight hitting it. It is normally mounted on the dashboard of the car behind the windscreen. In direct sunlight, you will feel hotter than the actual ambient cabin temperature due to the radiation produced by the sun's rays. It is the job of the sun load sensor to pick up this added solar radiation and adjust the loading on the climate control system to compensate, regardless of passenger compartment temperature.
Steering angle and torque sensor	A steering torque sensor is used to determine how much effort is being placed on the steering by the driver or road conditions. This information is then used to regulate the amount of power assistance given by the steering or help a stability control system to maintain handling. It normally consists of two Hall effect sensors mounted either side of a torsion bar fitted on the steering column. As the steering is turned the two signals produced are compared and depending on how far out of phase they are, this will give an indication of the effort/torque placed on the steering column. Another two types of steering angle sensor are variable resistance and optical. With a variable resistance, as the steering is turned from left to right, a voltage signal between 0 volts and 5 volts will be sent to the ECU which corresponds to steering position. Many vehicles produce a positive reading when turning to the right and a negative reading when turning to the left. An optical sensor uses a light emitting diode LED and a receiver that is interrupted by a 'chopper plate' which has small windows. As the steering is turned a digital signal is produced as the light from the LED is picked-up or blocked and compared with another sensor on the steering column.

Automotive Sensors & Waveform Analysis

ABS – Anti-lock Braking System.

TRC – Traction Control System.

EBD – Electronic Brake-force Distribution.

VSC – Vehicle Stability Control.

MRE – Magnetic Resistive Element.

Potentiometer – a form of variable resistor.

Drive-by-wire – a system where the mechanical cable has been removed, and the throttle butterfly is controlled by electronics and motors instead.

Depression – a low pressure or vacuum.

Thermistor – a thermal resistor.

Negative Temperature Coefficient NTC – a sensor in which resistance falls as temperature increases.

Positive Temperature Coefficient PTC – a sensor in which resistance rises as temperature increases.

Reluctor – a form of toothed pulse wheel used in conjunction with inductive speed sensors.

Transistors – miniature electronic switches with no moving parts.

Tolerance – an acceptable difference between an upper and lower limit.

Stratified – consisting of layers.

Catalyst – something that starts and maintains a chemical reaction, but is not directly affected itself.

Multiplexed – a system which can conduct multiple tasks at the same time.

Photoelectric – electronic effects produced by light.

Although an oscilloscope is designed to be a passive piece of electrical test equipment, and provide no potential to a circuit, it is <u>not recommended for use on supplementary restraint system SRS</u> sensors or actuators. An equipment fault or incorrect connection to a circuit could lead to accidental deployment of an airbag or seatbelt pretensioner, leading potential damage or personal injury.

Automotive Sensors & Waveform Analysis

Oscilloscope quick set up guide for use with sensors:

Measuring voltage:

How to:

Note: The oscilloscope probes may come in different colours, but for the sake of simplicity we will call them red and black here.

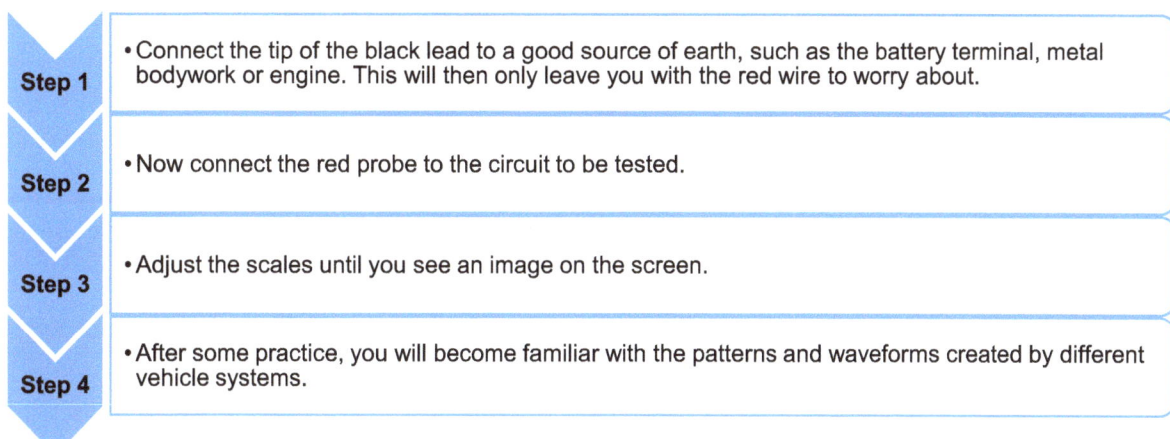

Step 1 • Connect the tip of the black lead to a good source of earth, such as the battery terminal, metal bodywork or engine. This will then only leave you with the red wire to worry about.

Step 2 • Now connect the red probe to the circuit to be tested.

Step 3 • Adjust the scales until you see an image on the screen.

Step 4 • After some practice, you will become familiar with the patterns and waveforms created by different vehicle systems.

Wheel speed sensors (inductive, MRE, Hall Effect)

Anti-lock braking systems, traction control and vehicle stability systems rely on accurate speed measurements coming from each wheel. Except for manoeuvring or cornering conditions, when road wheels will move on a different arc or curve, all four wheels should be traveling at roughly the same speed. If any wheel is found to be rotating at a significantly different speed than the others, the control system will assume that this relates to wheel slip caused by either acceleration or deceleration. When combined with signals from either the brake light or accelerator circuit, the ECU can then decide whether to intervene with ABS or traction control.

Inductive wheel speed sensor

An inductive wheel speed sensor is passive, as it takes information from its environment without external power from the ECU. It uses a change in a **magnetic flux** created by a rotating toothed wheel (known as a **reluctor**) to create an alternating voltage output which will be proportional to the speed of the wheel.

Magnetic flux – a measure of the quantity of magnetism when calculating the invisible field lines created in a specific area.

Reluctor – a rotating toothed wheel, used in conjunction with an inductive sensor.

Automotive Sensors & Waveform Analysis

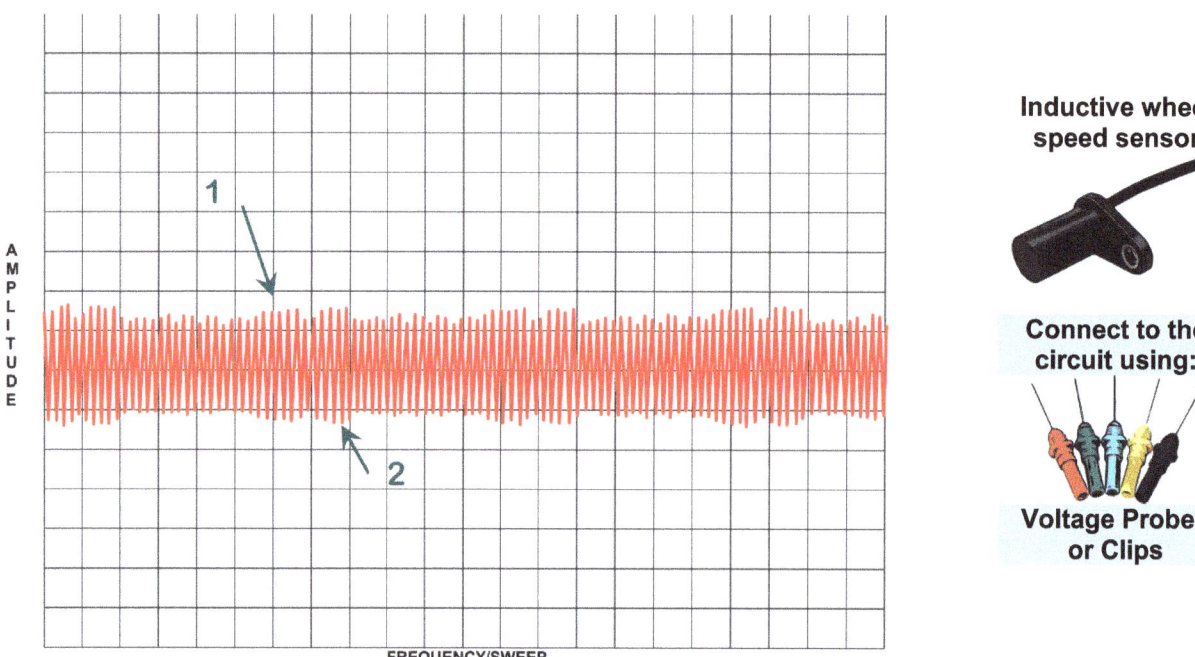

Figure 4.3 Inductive wheel speed sensor

Table 4.3 Waveform analysis inductive wheel speed sensor

Waveform component	Description
1	This shows the positive voltage produced by the inductive wheel speed sensor as the reluctor tooth moves towards it. Its height (or amplitude) will vary with speed. (i.e. it will get taller as the wheel speed increases, or shorter as the wheel speed slows down).
2	This shows the negative voltage produced by the inductive wheel speed sensor as the reluctor tooth moves away from it. The frequency of the waveform (how close together the alternating peaks and troughs form) will also alter with the speed of the wheel. (i.e. the wave will get closer together as the speed increases and further apart as the wheel slows down).

As an inductive wheel speed sensor generates its own electric voltage; it can be tested in-situ with the harness plug disconnected. By placing the oscilloscope probes on the two electrical connectors and spinning the wheel, a corresponding waveform should be produced.

This may however lead to misdiagnosis if the type of sensor is incorrectly identified. It is good practice to conduct the test by back-probing the sensor wires (possibly at the ECU harness), switch on the ignition and spin the wheel. In this way, it won't matter which type of sensor you are measuring, a waveform should be produced and can then be analysed.

Automotive Sensors
& Waveform Analysis

A disadvantage of inductive wheel speed sensors is their ability to produce electrical voltage at slow rotation speeds.
Due to their slow speed inaccuracy, many wheel speed sensors are of the Hall effect or MRE type.
Both of these sensor types are voltage fed (active), meaning they have the ability to produce signals at extremely low rotation speeds, making them more reliable and accurate.

Magnetic Resistive Element (MRE) wheel speed sensor

A Magnetic Resistive Element (MRE) wheel speed sensor is often similar in appearance to an inductive wheel speed sensor. It may also have only two wires in its harness plug, however, its construction and operation is very different. Instead of a toothed reluctor wheel, a magnetic pulse ring with alternating north and south magnetic poles is rotated past the sensor as the road wheel turns.
The sensor itself has an internal resistor that is sensitive to the change of strength in a magnetic field.
A reference voltage (often around 5 volts) is passed through the MRE sensor and, as the pulse ring is rotated past, a corresponding alternating voltage is created at the output wire and sent to the ECU for interpretation.

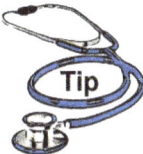

As an MRE sensor uses a reference voltage to generate a signal, it cannot be tested in-situ with the harness plug disconnected in the same way that an inductive sensor can.
It is good practice to conduct the test by back-probing the sensor wires (possibly at the ECU harness), switch on the ignition and spin the wheel. In this way, it won't matter which type of sensor you are measuring, a waveform should be produced and can then be analysed.

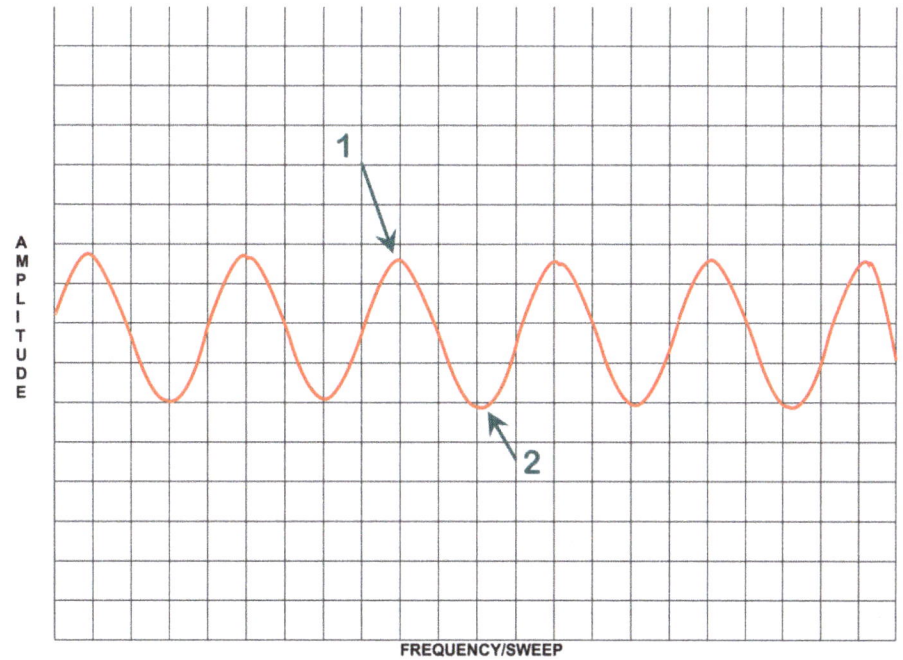

Figure 4.4 MRE wheel speed sensor

Automotive Sensors & Waveform Analysis

Table 4.4 Waveform analysis MRE wheel speed sensor

Waveform component	Description
1	This shows an increasing voltage produced by the MRE wheel speed sensor as its internal resistance falls due to the effect of the magnetic flux created by the pulse wheel. Unlike the inductive sensor, where an MRE is supplied with a reference voltage, its height (or amplitude) will not increase or decrease due to wheel speed. (i.e. no matter how fast or slow the wheel rotates, the waveform will remain the same height).
2	This shows a reducing voltage produced by the MRE wheel speed sensor as its internal resistance increases due to the effect of the magnetic flux created by the pulse wheel. The frequency of the waveform (how close together the alternating peaks and troughs form) will alter with the speed of the wheel. (i.e. the wave will get closer together as the speed increases, and further apart as the wheel slows down; an MRE can provide a signal at very low rotation speeds).

To help ensure the integrity of the magnetic pulse ring located in the wheel hub, viewing tools are available for diagnostic purposes. They consist of a clear laminated plastic window with a magnetic sensitive liquid sandwiched between the layers. When held over the pulse ring, the liquid is distorted to show a pattern of magnetic fields and will help identify any faulty portions.

Hall effect wheel speed sensor

A Hall effect wheel speed sensor creates a digital signal which will be shown on an oscilloscope as a distinctive square waveform. It is mounted on a wheel hub or driveshaft where a slotted ring can rotate and provide the reference for wheel speed. As the slotted plate moves past the wheel speed sensor, a small magnetic field is disrupted inside the Hall effect senor, switching the circuit on and off, providing a digital signal proportional to wheel speed for the ECU to process. The supplied reference voltage will vary from manufacturer to manufacturer, however, as this sensor provides a digital signal which needs no conversion by the ECU, its amplitude is far less important to the calculation of wheel speed.

Automotive Sensors & Waveform Analysis

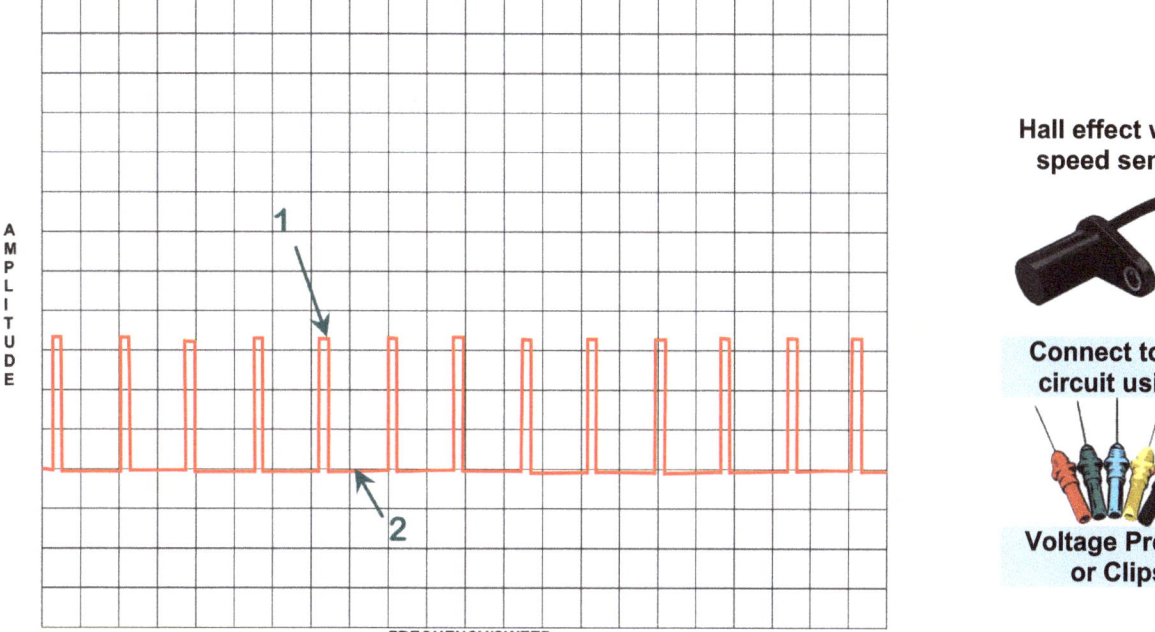

Figure 4.5 Hall effect wheel speed sensor

Table 4.5 Waveform analysis Hall effect wheel speed sensor

Waveform component	Description
1	This point represents voltage switched-on by the Hall effect sensor and will not change in height (amplitude) as wheel speed changes.
2	This point represents voltage switched-off by the Hall effect sensor and will change in frequency as wheel speed changes. (i.e. the wave will get closer together as the speed increases and further apart as the wheel slows down; a Hall effect sensor can provide a signal at very low rotation speeds).

As Hall effect sensors use a reference voltage in order to generate a signal, they cannot be tested in-situ with the harness plug disconnected in the same way that an in inductive sensor can.
It is good practice to conduct the test by back-probing the sensor wires (possibly at the ECU harness), switch on the ignition and spin the wheel. In this way, it won't matter which type of sensor you are measuring, a waveform should be produced and can be analysed.

Automotive Sensors & Waveform Analysis

Throttle pedal sensors (analogue and digital)

As most vehicles are now drive-by-wire (i.e. no throttle cable, but instead controlled by motors and electronics) the position of the accelerator pedal is vital for the ECU to calculate driver demands.

For safety reasons, most accelerator pedal position sensors are multi-track potentiometers, providing back-up signals that can be compared by the ECU to ensure the reliability of the information provided. As the throttle pedal is pressed by the driver, an analogue signal is produced, showing a rising and falling voltage which is proportional to the position of the pedal and will react in accordance with driver inputs (even showing snap open and close of the throttle).

Analogue/Digital pedal position sensors are also used by some manufacturers. These sensors provide an analogue signal from a potentiometer, but on the second channel of an oscilloscope, a square waveform will be produced with a frequency that is proportional to the position of the pedal at the time.

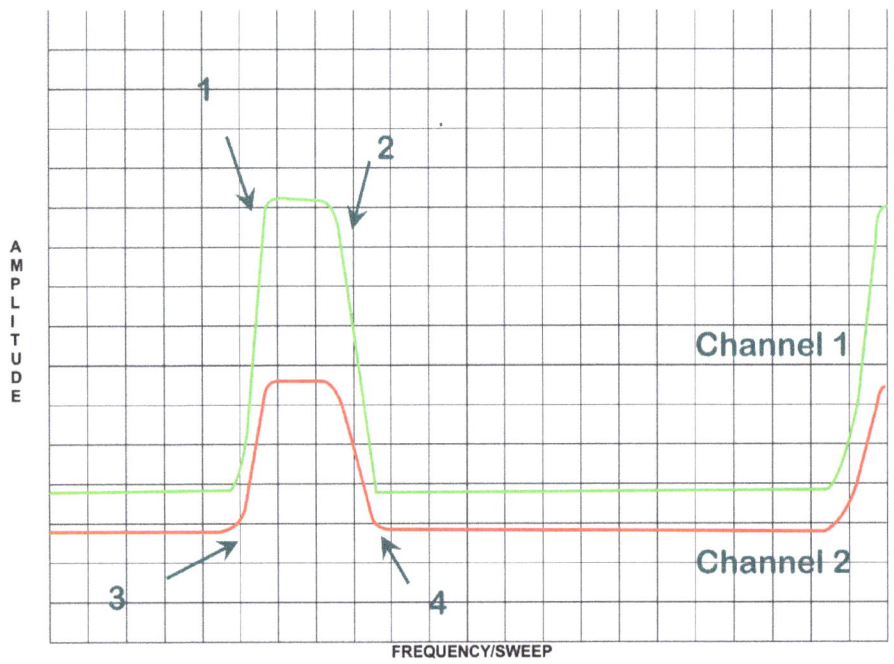

Figure 4.6 Analogue throttle pedal sensor

Automotive Sensors & Waveform Analysis

Table 4.6 Waveform analysis throttle pedal position sensor (analogue)

Waveform component	Description
1	Channel 1. This shows a rising voltage created by the pedal potentiometer measured on channel 1. This rising voltage is caused by a reduction in resistance within the potentiometer as the throttle is pressed.
2	Channel 1. This shows a falling voltage created by the pedal potentiometer measured on channel 1. This falling voltage is caused by an increase in resistance within the potentiometer as the throttle is released.
3	Channel 2. This shows a rising voltage created by the pedal potentiometer measured on channel 2. This rising voltage is caused by a reduction in resistance within the potentiometer as the throttle is pressed.
4	Channel 2. This shows a falling voltage created by the pedal potentiometer measured on channel 2. This falling voltage is caused by an increase in resistance within the potentiometer as the throttle is released.

The amplitude (height) of the waveforms will be dependent on the supply voltages provided by the manufacturers and the internal resistance of the potentiometers. It is not unusual to have slightly different amplitudes between the two potentiometer tracks as this can help the ECU determine which potentiometer is being read. Always check the manufacturer specifications before assuming that a fault is occurring.

Vane air flow meter

In order for the engine management ECU to calculate the correct air/fuel ratio, it needs to know the amount of air entering the engine. There are several different measurement devices used to perform this function, and the type used on some early fuel injection systems was the vane air flow sensor. In this type of air measurement device, a sprung loaded internal flap is pushed open by the movement of the incoming air. Connected to this flap is a potentiometer which will produce an output voltage that corresponds to its movement and can be interpreted as air flow.

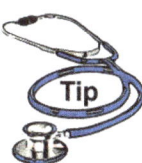

It is possible to test a vane type air flow sensor without the engine running. With the induction hose leading to the sensor disconnected, and the ignition switched on, the flap can be manually operated with the tip of a screwdriver, and the resulting waveform analysed.

Automotive Sensors & Waveform Analysis

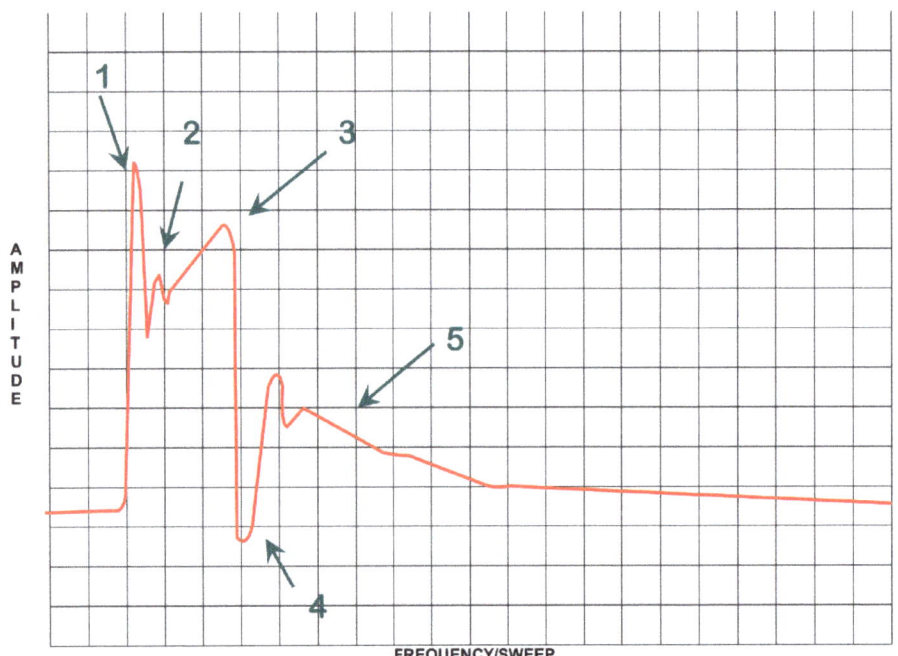

Figure 4.7 Vane type air flow sensor

Table 4.7 Waveform analysis vane type air flow sensor

Waveform component	Description
1	From a small initial voltage of around 1 volt at idle, as the throttle is opened, the internal flap of the airflow meter moves and the voltage rapidly rises to give an initial peak. This initial spike on the waveform is caused by the **inertia** of the moving flap.
2	Following the initial peak shown on the waveform, caused by the inertia, the voltage momentarily drops as air flow settles against the internals of the sensor, before starting to rise again.
3	Once the sensor has settled into a steady airflow, the voltage will rise until it reaches a good approximation of the amount of air entering the engine. (The amplitude will depend on how hard the engine is revved during the test).
4	When the throttle is released, the flap inside the airflow meter closes, and once again inertia keeps it moving, but this time in the opposite direction. This will cause the voltage to momentarily drop below an idle voltage value.
5	As the engine management systems tries to even out the RPM using an idle speed control valve, there is another slight rise in voltage before the pattern returns to its initial idle position.

Inertia – the tendency for a body to resist acceleration or change of direction.

Automotive Sensors & Waveform Analysis

Notes: It is not unusual to see a rough image with interference caused by induction pulses as they act on the flap inside a vane type airflow meter. However, the vane type air flow meter contains a compensation flap, to reduce these pulsations as much as possible.

Mass air flow meter

In order for the engine management ECU to calculate the correct air/fuel ratio, it needs to know the amount of air entering the engine. There are several different measurement devices used to perform this function, and a relatively common type is the mass air flow meter.

This type of sensor uses a wire or metallic/ceramic element which is heated by electric current supplied from a system relay. As air passes over the heated element it is cooled and requires more current to bring it back to its set operating temperature. The amount of current flowing is proportional to the mass of the air entering the intake system.

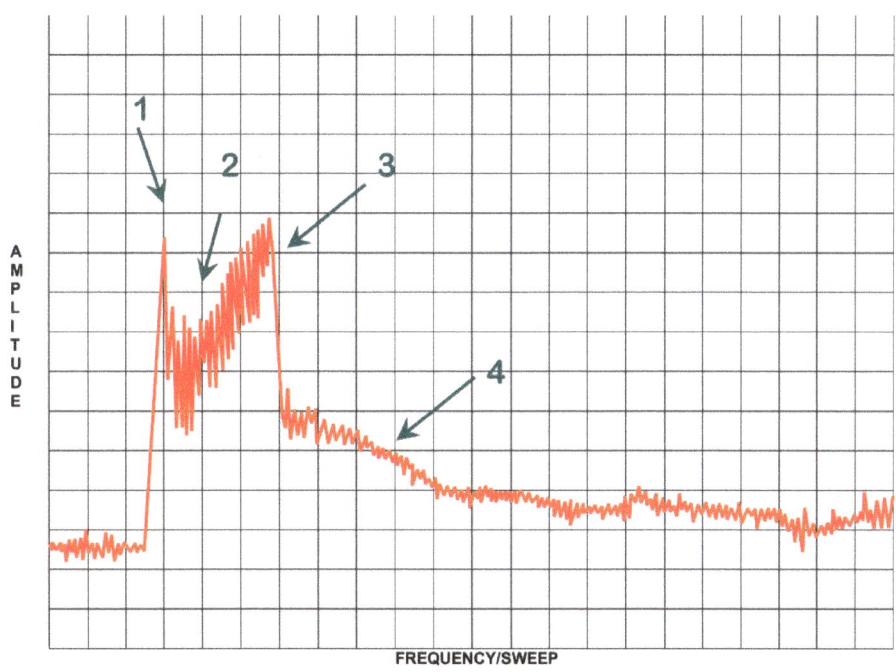

Figure 4.8 Mass air flow sensor

Automotive Sensors & Waveform Analysis

Table 4.8 Waveform analysis mass air flow sensor

Waveform component	Description
1	From a small initial voltage of around 1 volt at idle, as the throttle butterfly is opened, air is drawn into the intake system and there is a rapid rise in voltage.
2	After an initial peak of voltage, caused by the inertia of air rushing into the intake system, the voltage momentarily drops as air flow settles inside the sensor, before starting to rise again. Once the engine has settled into a stable running condition, the voltage rises until it reaches a good approximation of how much air is entering the intake system. (The amplitude will depend on how hard the engine is revved during the test).
3	When the throttle butterfly is released, the air flow initially falls quite rapidly, until the engine management system intervenes to prevent the engine RPM dropping too low and stalling.
4	As the engine management system tries to even out the RPM using an idle speed control valve, there is another slight rise in voltage before the pattern returns to its initial idle position.

It is sometimes possible to detect exhaust gas recirculation (EGR) within the waveform of the mass air flow meter. This is because as EGR occurs, air flow in the intake will reduce slightly. If a second channel is connected to the EGR actuator, and the vehicle is driven, the waves from the two displays can be compared and you will be able to confirm that exhaust flow is taking place. This test is a good method of checking for physical exhaust gas flow which could be prevented by blockage or failure of the EGR valve.

Diesel air flow meter

Unlike a petrol engine, the air flow meter of a Diesel engine is not directly responsible for adapting the fueling to achieve the correct air/fuel ratio. A Diesel engine does not require a throttle butterfly for engine speed control, but when combined with readings from an air flow meter, can regulate **EGR**, reducing **NOx** emissions and the amount of oxygen reaching the cylinder; this will then help regulate air/fuel ratio for different running conditions.
Although many manufacturers are using a hot wire mass air flow meter, similar to those found on petrol engines, its signature waveform is slightly different in shape.

EGR – exhaust gas recirculation is a method used to introduce exhaust gas into the intake, which helps to reduce the production of the pollutant NOx.

NOx – oxides of nitrogen; a harmful pollutant created during the combustion process, when temperatures exceed 1,800 degrees Celsius.

Automotive Sensors & Waveform Analysis

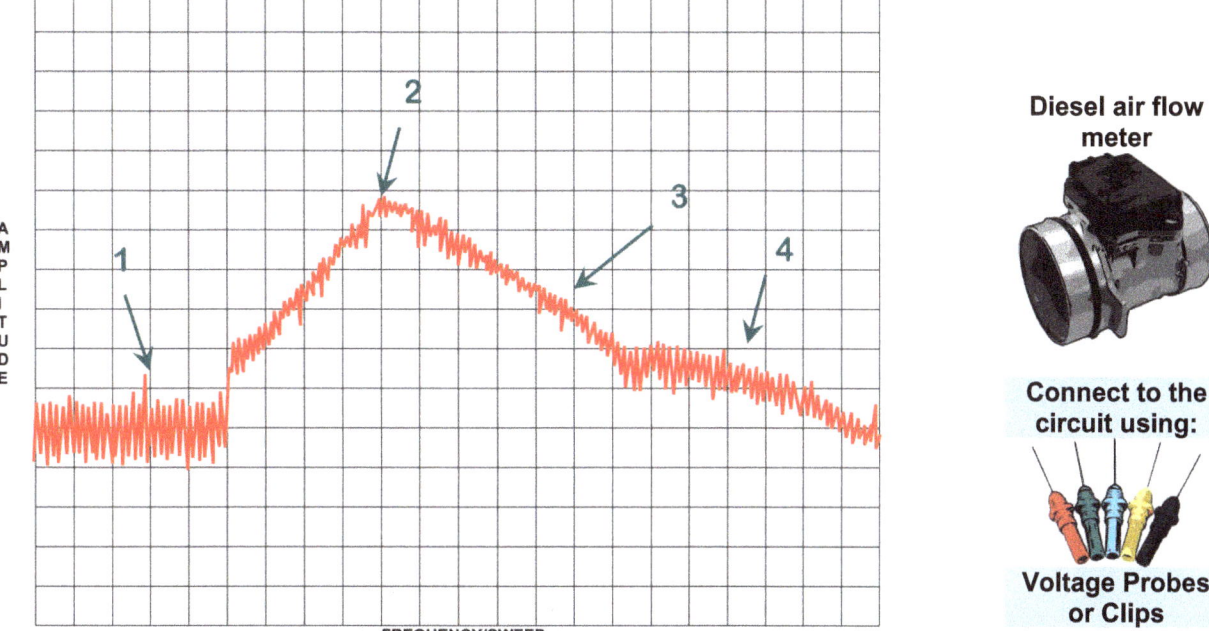

Figure 4.9 Diesel air flow meter

Table 4.9 Waveform analysis Diesel mass air flow sensor

Waveform component	Description
1	This section of the waveform represents the voltage signal at idle. (The interference shown in the pattern is caused by the induction strokes of the engine).
2	This section shows a steady rise to a peak caused by the engine revs increasing; with a mixture of air flow and exhaust gas recirculation through the intake. The peak amplitude will depend on the speed and load of the engine at the time of measuring.
3	This section shows a reducing air flow as the engine revs fall when the accelerator pedal is released.
4	If an idle speed control valve is fitted (regulating the quantity of fuel injected), you may see a gradual return to idle as the engine management system operates in anti-stall.

Vortex air flow meter

The vortex sensor can take a very accurate, digital measurement of airflow entering the intake system of a petrol engine. This type of air flow sensor is relatively rare, but may sometimes be confused with a hot wire mass air flow sensor due to its external appearance. If the pipe connecting it to the intake system is removed, it can sometimes be identified by a diffuser and a triangular shaped column mounted vertically in the airstream. As air passes over this column it is split into two swirling whirlpools or vortices that turn in opposite directions. In the vortex sensing chamber, the disruption to the airflow caused by the swirling air currents is detected by a pressure or sonic pick-up and converted into a digital signal.

Automotive Sensors & Waveform Analysis

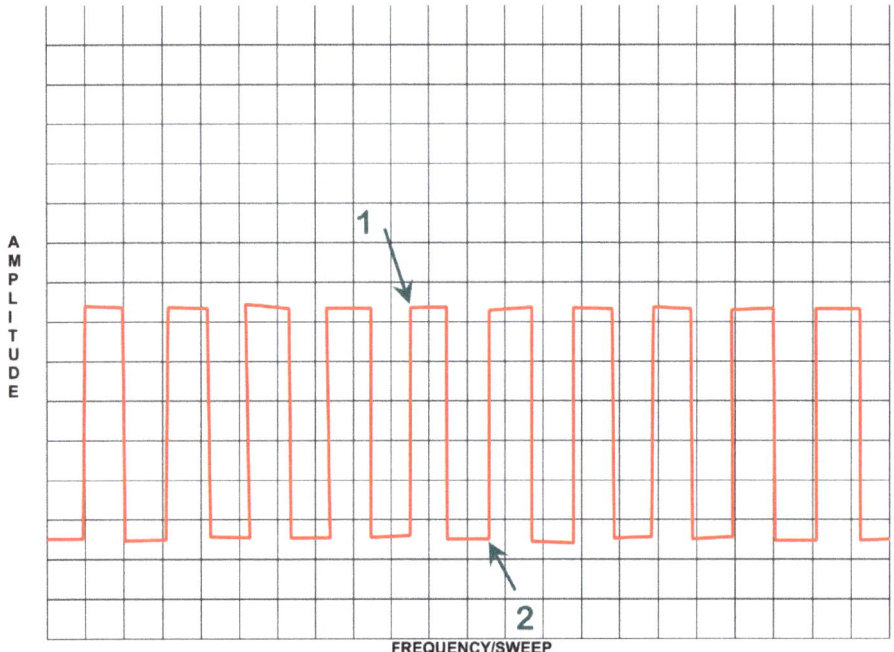

Figure 4.10 Vortex air flow meter

Table 4.10 Waveform analysis vortex air flow sensor

Waveform component	Description
1	This shows the on period of the of the digital signal produced by the vortex generator sensor. As engine speed increases and decreases its amplitude should remain steady.
2	This shows the off period of the digital signal produced by the vortex generator sensor. As engine speed increases and decreases, so will the frequency of the waveform.

Manifold absolute pressure sensor (analogue)

The analogue manifold absolute pressure sensor (MAP) can be mounted on or near the intake manifold, after the throttle butterfly, often into the **plenum chamber**. It measures the vacuum (or depression) in the manifold and allows the engine management ECU to calculate the amount of fuel to inject in a similar manner to an air flow meter. The MAP sensor is also used by the engine management ECU to help determine the load placed on the vehicle due to driving conditions, and allows it to make changes to ignition timing advance/retard.

Plenum Chamber – an air reservoir normally formed in the intake manifold design.

Automotive Sensors & Waveform Analysis

Notes: Some early manifold pressure sensors were constructed as an integral part of the engine management ECU mounted within the engine bay. Due to the lack of access to the output voltages from this type of MAP, it may not be possible to check its signal using an oscilloscope. If this is the case, voltages should be checked using live data from a scan tool.

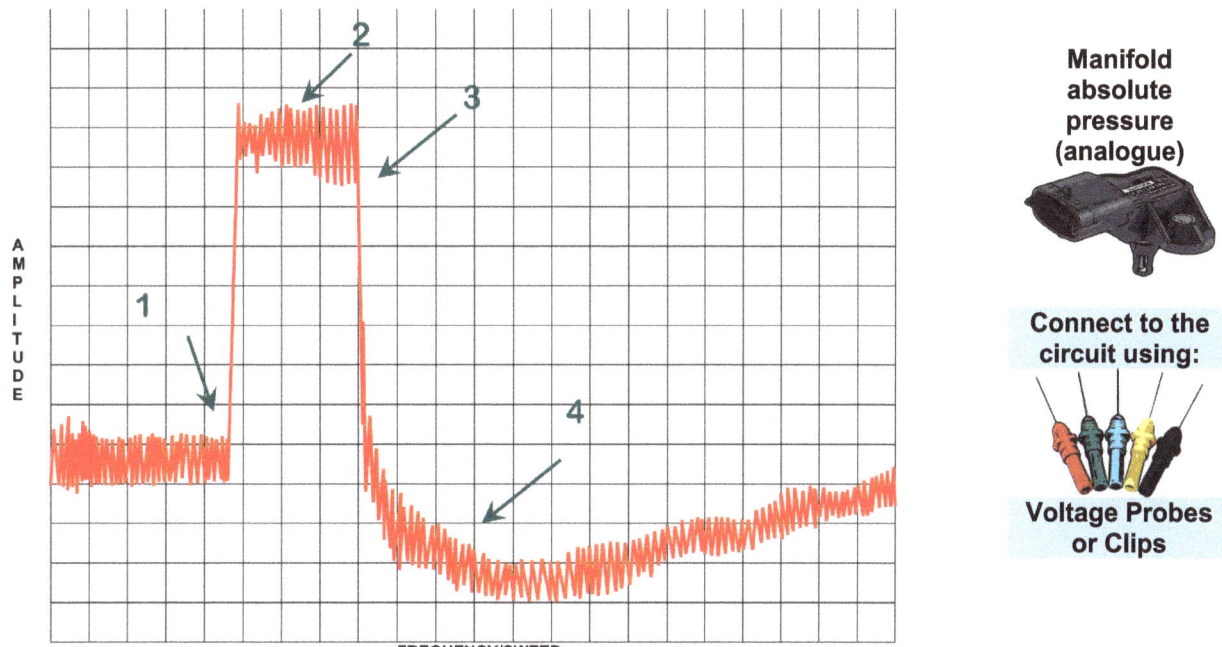

Figure 4.11 Analogue manifold absolute pressure sensor (MAP)

Table 4.11 Waveform analysis analogue manifold absolute pressure sensor

Waveform component	Description
1	This is where the throttle butterfly is opened and vacuum in the intake manifold rapidly drops causing a rise in voltage shown on the waveform.
2	With the throttle held open and vacuum low in the manifold, voltage remains high. The interference that can be seen on the waveform is caused by the engine intake pulsations of the pistons.
3	When the throttle is released, the voltage rapidly falls as vacuum in the manifold rises once again, caused by the restriction to the incoming air from the throttle butterfly.
4	At this point on the waveform, the voltage momentarily falls below idle values as the engine draws extra vacuum caused by the snapped closed throttle butterfly. As the revs settle, the voltage returns to idle values.

Automotive Sensors & Waveform Analysis

> If the voltage output is lower than expected from a manifold pressure sensor, the engine will run rich as the ECU assumes that there is more air flow. This condition, however, could be caused by a damaged vacuum connection to the intake manifold, or worn engine components reducing induction performance.
> A less likely outcome is that the voltage output is constantly higher than expected, making the engine run weak. This condition could be caused by a restriction in the intake system causing a higher than expected depression in the manifold.
> As many manufacturers use similar voltages for the operation of a manifold absolute pressure sensor, it should be possible to compare output values to another car in order to check your diagnosis.

Air intake pressure sensor

An air intake pressure sensor is used by the engine management system to help monitor boost pressure from a turbo charger. It will work in a similar manner to a manifold absolute pressure sensor, but the voltage on the waveform will rise and fall proportionately with boost pressure.

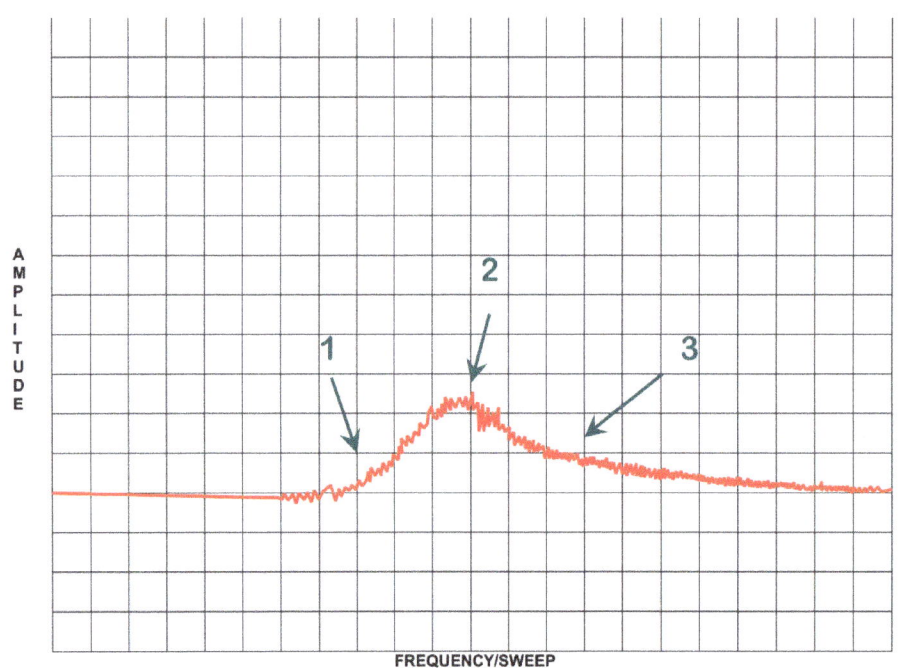

Figure 4.12 Air intake pressure sensor

Table 4.12 Waveform analysis air intake pressure sensor

Waveform component	Description
1	This point on the waveform shows the pressure in the intake system starting to rise as the turbo charger starts to boost.

Automotive Sensors & Waveform Analysis

Table 4.12 Waveform analysis air intake pressure sensor

2	This point on the waveform represents the maximum boost created by the turbo charger. The interference that might be seen on the waveform is created by the intake pulses of the engine on its induction strokes.
3	This point on the waveform represents the pressure falling as the turbo charger slows down and boost falls.

Manifold absolute pressure sensor (digital)

The digital manifold absolute pressure sensor (MAP) is used for the same purpose and works in the same manner as the analogue version, however, it has its own integrated circuit which processes the signals to digital as part of its construction. The advantage of having the sensor convert its signal to digital means that the data can be freely shared by other vehicle systems and the engine management control unit has less work to do as the information is ready for processing. Unlike the analogue version, a square waveform is produced with a constant amplitude, but a frequency that changes in relation to manifold pressure.

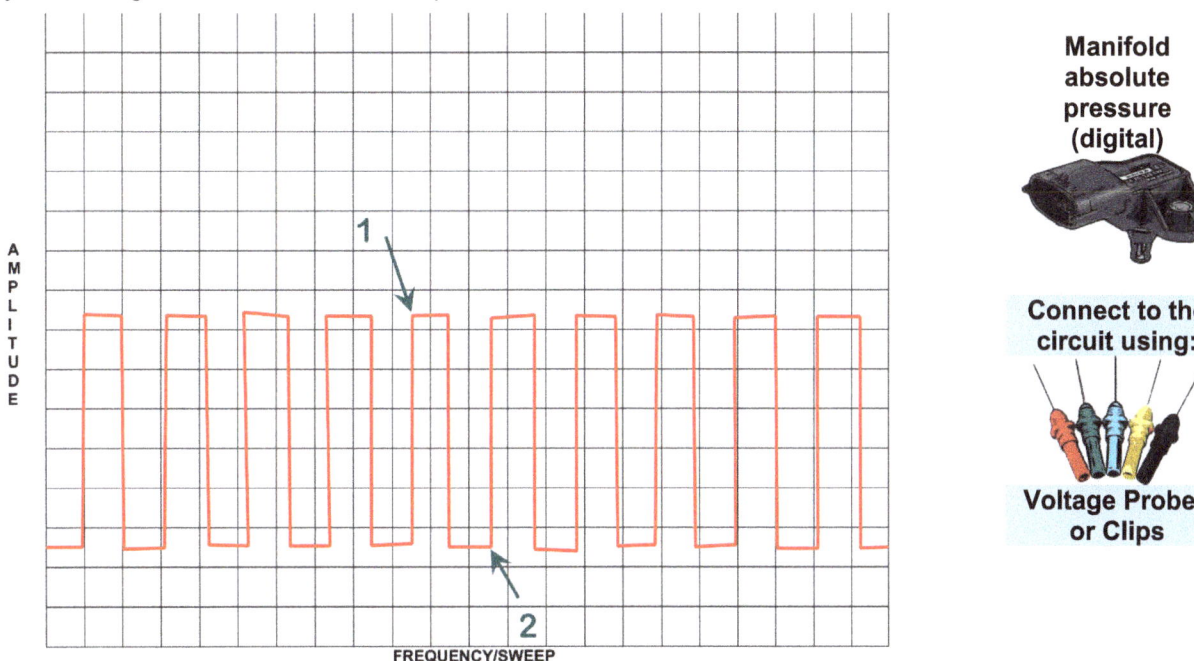

Figure 4.13 Digital manifold absolute pressure sensor (MAP)

Table 4.13 Waveform analysis digital manifold absolute pressure sensor

Waveform component	Description
1	This shows the on period of the of the digital signal produced by the manifold absolute pressure sensor. As engine speed increases and decreases its amplitude should remain steady.
2	This shows the off period of the digital signal produced by the manifold absolute pressure sensor. As engine speed increases and decreases, so will the frequency of the waveform.

Automotive Sensors & Waveform Analysis

Crankshaft sensor (inductive)

An inductive crankshaft sensor uses the change in magnetic flux, produced by a rotating toothed wheel (known as a reluctor), to create an alternating voltage output which will be proportional to the speed of the engine. Design of the reluctor wheel will vary from manufacturer to manufacturer and dictate the shape of the waveform produced. The example shown in this description is typical, however, care should be taken when conducting a diagnosis to allow for any differences in manufacturer design.

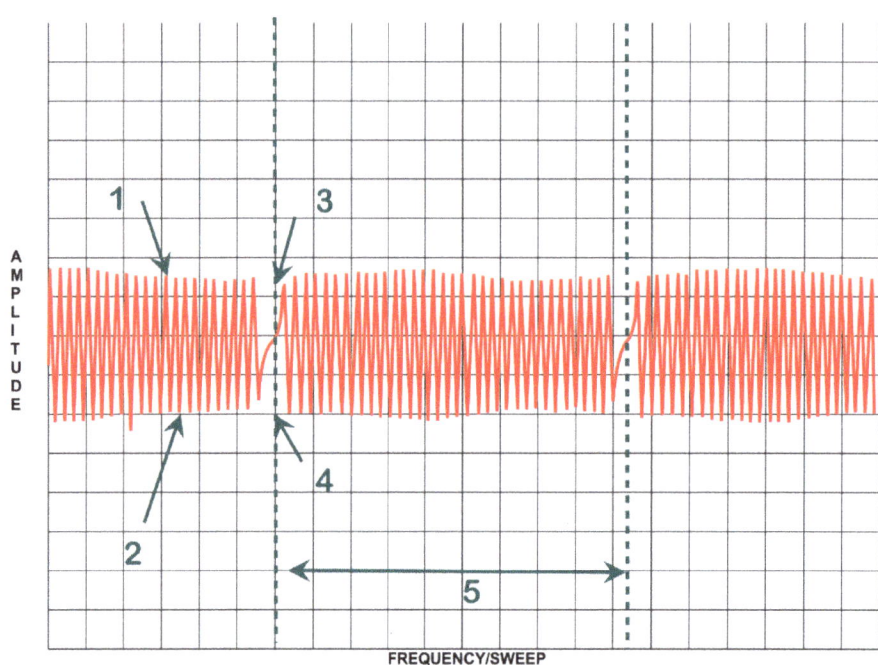

Figure 4.14 Inductive crankshaft sensor

Table 4.14 Waveform analysis inductive crankshaft sensor

Waveform component	Description
1	This shows a positive voltage being generated by the inductive sensor as the reluctor tooth rotates towards it. The amplitude or height of this signal will increase with speed and reduce as the engine slows down.
2	This shows a negative voltage being generated by the inductive sensor as the reluctor tooth rotates away from it. The amplitude or height of this signal will vary with engine speed in a similar manner to waveform component 1.
3	The change in the pattern at this point shows the location where teeth have been removed from the reluctor wheel, and helps the ECU determine the position of the crankshaft. This may indicate "Top Dead Centre", however, manufacturers are at liberty to use this to indicate a number of different positions such as, all pistons in line halfway up the bore, or any number of degrees before TDC. It is important to check with manufacturers specifications before making any diagnostic decisions based on this position.

Automotive Sensors & Waveform Analysis

Table 4.14 Waveform analysis inductive crankshaft sensor

4	This shows a cursor placed at the midpoint of the wave where reluctor teeth have been removed. This can help to identify the starting point of one crankshaft revolution.
5	By placing another cursor at the next corresponding point on the waveform and measuring the distance between the two points, it is possible to show the speed of one complete crankshaft revolution.

By placing cursor lines on the oscilloscope screen at the corresponding positions where the teeth are missing from the reluctor wheel, some oscilloscopes are able to calculate the RPM of the crankshaft.
Having set the cursors, the frequency (measured in Hertz Hz) and the Delta Δ (the time between the cursor lines) is displayed.
If RPM is not automatically calculated, this can be worked out using the Delta Δ.
To calculate the RPM of the crankshaft from the waveform if the timescale is set to milliseconds (m/s):
Take the Delta Δ time displayed and perform the following calculation:
60,000 ÷ Delta Δ = crankshaft RPM

Crankshaft sensor (Hall Effect)

A Hall effect crankshaft sensor develops a magnetic field which is interrupted by a rotating slotted disc. As the engine turns, the slots in the disc create an on and off signal at the Hall effect sensor, producing a digital/square pattern which can be used by the engine management system to show RPM and crankshaft position. The design and location of the sensor may vary from manufacturer to manufacturer, but its operation will include a section that indicates 'top dead centre' or other crankshaft position.

The sensor will have three electrical connections:
- Feed
- Signal
- Ground

An advantage of this type of sensor design is that it does not require extra electronics for conversion to digital before it can be used by an electronic control unit ECU.

Automotive Sensors & Waveform Analysis

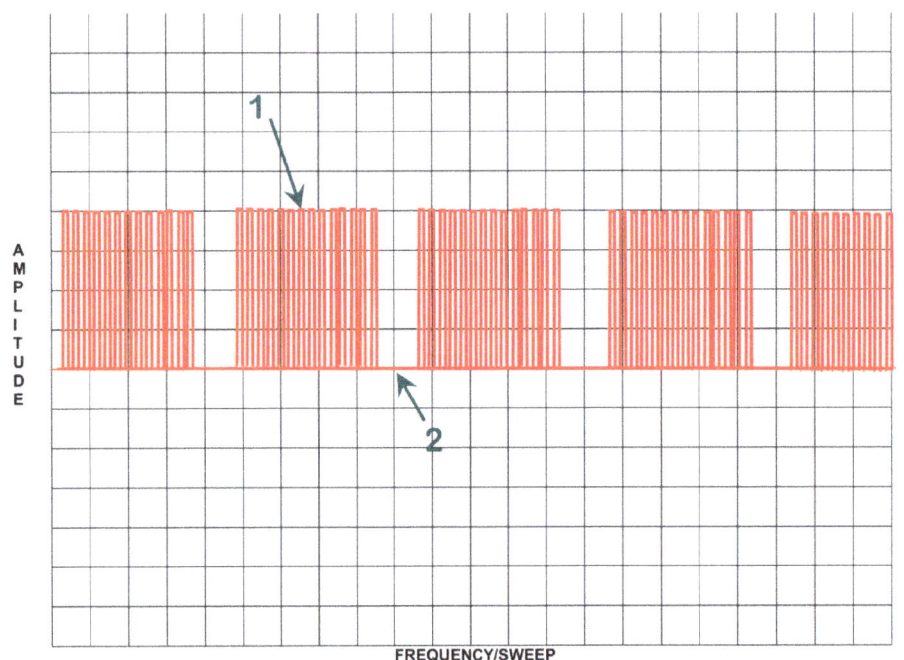

Figure 4.15 Hall effect crankshaft sensor

Table 4.15 Waveform analysis Hall effect crankshaft sensor

Waveform component	Description
1	This point on the waveform shows the digital signal that is produced as the slotted disc rotates past the Hall effect sensor. As the engine speeds up and slows down, the frequency of the waveform will increase and decrease, but its amplitude will remain constant.
2	This section shows the interruption to the signal, indicating TDC (or other crankshaft position depending on manufacturer design).

Coolant temperature sensor

The engine coolant temperature (ECT) or coolant temperature sensor (CTS) is a thermal resistor or **'thermistor'**, where the internal resistance will be affected by temperature change. It is often a two-wire device that is located in the cylinder head or block, where the tip can be exposed to engine coolant.
Two main types of ECT are used in engine management systems, **negative temperature coefficient (NCT)** and **positive temperature coefficient (PCT)**.
The most common type used is an NTC, and this is shown in figure 4.16; its operation is described in table 4.16.

Automotive Sensors & Waveform Analysis

Thermistor – a temperature sensitive resistor (thermal resistor).

NTC – negative temperature coefficient. When used with thermistors, as temperature rises, resistance falls.

PTC – positive temperature coefficient. When used with thermistors, as temperature rises, resistance rises.

NTC and PTC coolant temperature sensors are very similar in external appearance. Care must be taken when testing or replacing, that you are aware of which type is being used.
With an NTC, resistance will fall as temperature rises.
With a PTC, resistance will increase as temperature rises.
(waveforms will be similar but move in opposite directions)

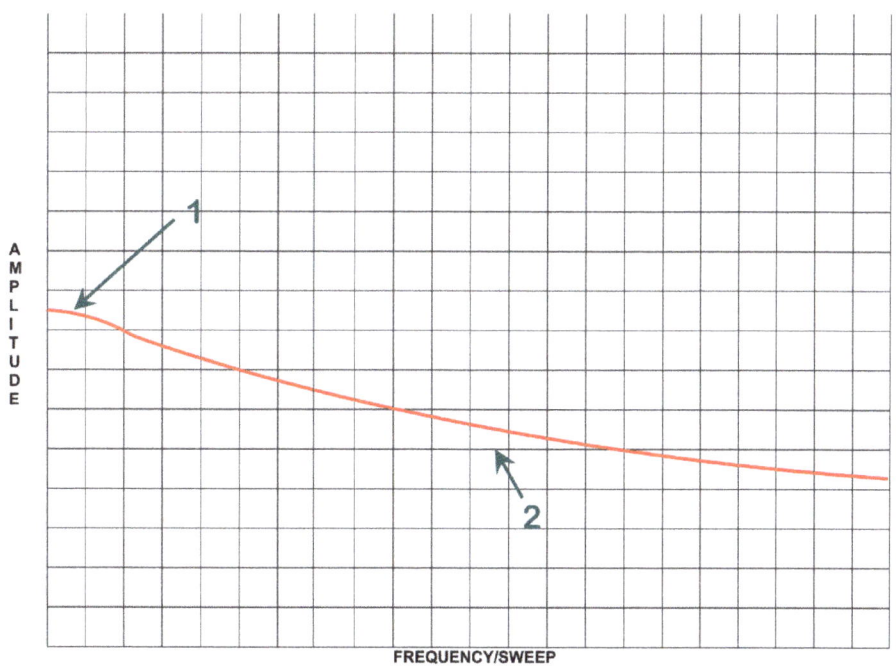

Figure 4.16 Engine coolant temperature sensor

Automotive Sensors & Waveform Analysis

Table 4.16 Waveform analysis coolant temperature sensor (NTC)

Waveform component	Description
1	Voltage is supplied by the engine management ECU and when the engine is cold, internal resistance will be high, meaning that the voltage on the signal wire will remain close to the input voltage.
2	With a very slow timescale selected (over a period of minutes), the engine started and warming up, the voltage will fall gradually on the signal wire as the internal resistance of the sensor reduces relative to temperature increase. The falling voltage pattern shown is proportional to the temperature of the coolant, and should be smooth with no sudden jumps or falls. If positive or negative spikes are shown within the waveform, this would be an indication of short or open circuit within the system.

Many coolant temperature sensors look externally similar, but can operate differently (NTC or PTC). There may often be no way to know which one is fitted unless you test it.
The electrical connections of some ECT's are colour coded and if different coloured sensors are fitted or supplied, you should ensure that they are the correct type for the vehicle.

Air temperature sensor

An air temperature sensor is a thermal resistor which operates in a similar manner to an engine coolant temperature sensor, and is used to help determine the density of air entering the engine. As with other types of temperature sensor, these can be NTC or PTC (see coolant temperature sensor).
Air temperature sensors can be a self-contained component, or it may form part of another sensor; Mass Air Flow (MAF) for example.
Its location should be considered carefully before testing as this will influence the types of diagnostic reading obtained. For example, the temperature of intake air will be considerably higher if it is mounted after a turbocharger.

When tested, most live air temperature readings will remain fairly constant. This means that in order to get useful diagnostic information, an air temperature change will have to be simulated by blowing hot or cold air across the sensor itself.

Automotive Sensors & Waveform Analysis

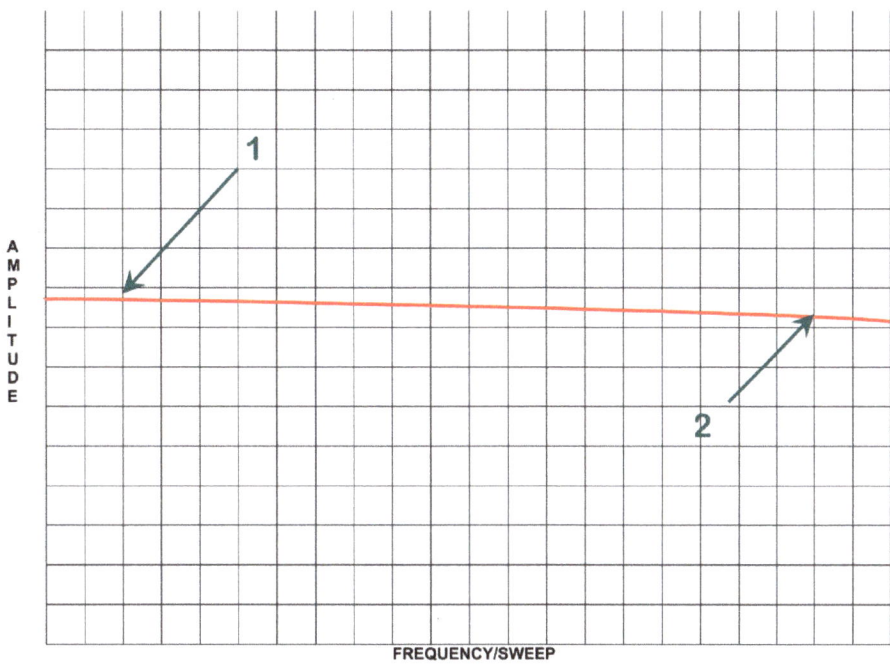

Figure 4.17 Air temperature sensor

Table 4.17 Waveform analysis air temperature sensor (PTC)

Waveform component	Description
1	The measured voltage at this point of the waveform is in direct proportion to the air temperature being measured.
2	If an air temperature change is simulated, the waveform should fall or rise in response. (This shows the effect of cooling the air on a PTC sensor). NTC – the wave will fall as temperature rises PTC – the wave will rise with an increase in temperature

Barometric pressure sensor

A barometric pressure sensor is used on some vehicles to help engine management determine the atmospheric pressure and therefore more accurately calculate the density of the air entering the intake system. This will then ensure that the correct air/fuel ratio is maintained regardless of altitude or weather conditions.

Because a barometric pressure sensor provides a constant voltage proportional to the atmospheric pressure at the time of capture, the waveform produced on an oscilloscope will remain flat and difficult to diagnose.
If a diagnostic trouble code is triggered relating to the sensor, it may be possible to test its operation by sealing the intake system around it and simulating a pressure change using a hand operated pressure/vacuum pump.

Automotive Sensors & Waveform Analysis

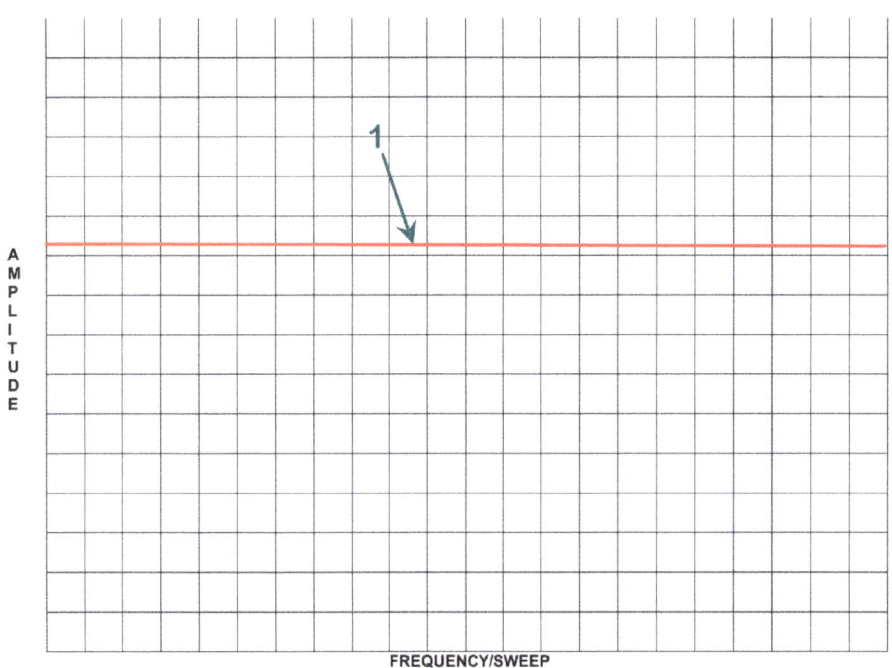

Figure 4.18 Barometric pressure sensor

Table 4.18 Waveform analysis barometric pressure sensor

Waveform component	Description
1	The waveform illustrated shows a continual voltage that is representative of the atmospheric pressure seen at the time of capture.

Camshaft sensors

The camshaft sensor on many vehicles is used as a reference signal to enable the engine management system to determine which cylinder is on its compression stroke, and therefore, control ignition timing and sequential fuel injection. This is often achieved by comparing the camshaft signal with the crankshaft position sensor signal. The sensor can be an inductive or Hall effect, and which type is used will depend on the manufacturer.

Hall effect camshaft sensor

A Hall effect camshaft sensor can often be identified by its three-wire connection. There will be a feed wire, an earth wire and a signal wire, which during operation will produce a digital square waveform that corresponds to camshaft position. When used in conjunction with the crankshaft signal, the engine management system can calculate correct ignition timing and sequential fuel injection.

Automotive Sensors & Waveform Analysis

The distance between two identical switching points on the waveform produced by a Hall effect camshaft position sensor normally represents 720 degrees of crankshaft revolution. (i.e. two complete rotations of the crankshaft).

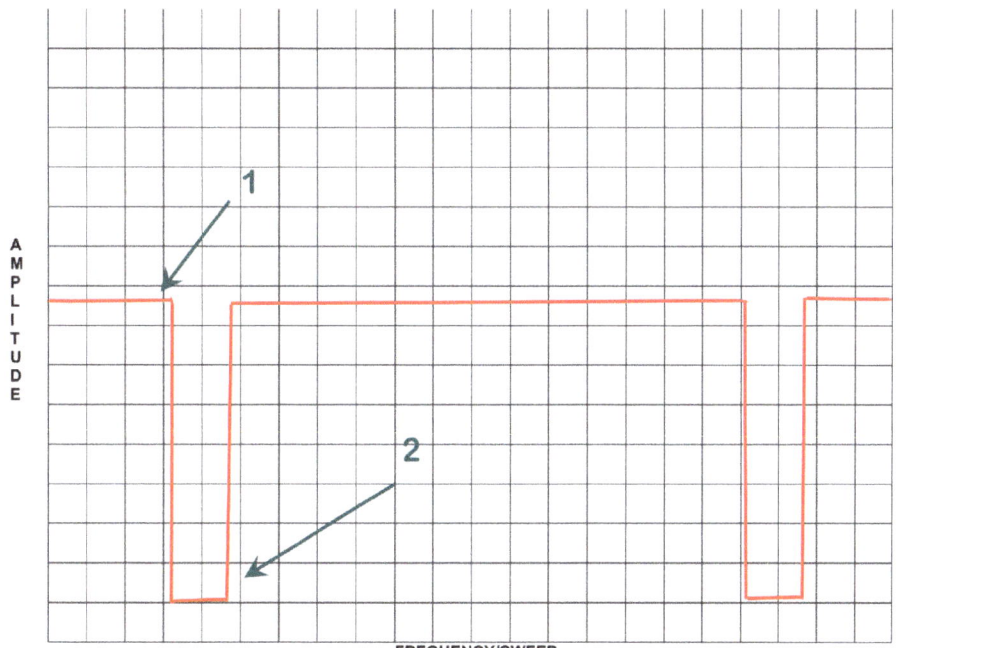

Figure 4.19 Hall effect camshaft sensor

Table 4.19 Waveform analysis Hall effect camshaft sensor

Waveform component	Description
1	This position on the waveform is produced when the signal wire receives the full system fed voltage from the ECU, and represents the off condition.
2	This point on the waveform shows where the voltage has been interrupted by the camshaft rotor section and the voltage should fall close to zero. This represents the on condition and relates to the exact position of the camshaft. A characteristic of a good Hall effect sensor is a clean, sharp, square waveform.

The engine management system of most vehicle manufacturers will need to see an effective signal from the camshaft sensor during cranking for start-up to confirm effective operation. If this signal is not seen, then the system may enter limited operating strategy (or 'limp-home').

Automotive Sensors & Waveform Analysis

Inductive camshaft sensor

The inductive camshaft sensor generates its own voltage signal when the magnetic field created by a permanent magnet is disrupted by the movement of a reluctor tooth. As the camshaft rotates, and the magnetic field is moved, a small voltage is generated inside a coil within the sensor and produces a characteristic sinewave when connected to an oscilloscope.

Each waveform pulse represents one complete turn of the camshaft and two complete turns of the crankshaft.

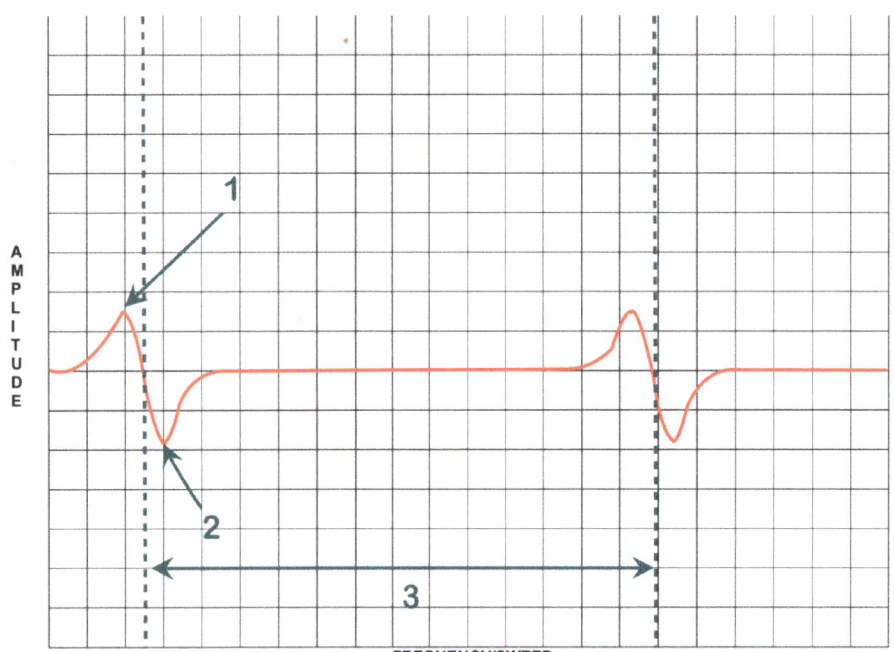

Figure 4.20 Inductive camshaft sensor

Table 4.20 Waveform analysis inductive camshaft sensor

Waveform component	Description
1	This is the point on the waveform, where the reluctor is moving towards the camshaft sensor, generating a positive voltage. The amplitude or height of this section will be proportional to engine speed and strength of magnetic field. The faster the engine runs or the stronger the magnetic field, the taller this point will be.
2	This is the point on the waveform, where the reluctor is moving away from the camshaft sensor, generating a negative voltage. The amplitude or height of this section will be proportional to engine speed and strength of magnetic field. The faster the engine runs or the stronger the magnetic field, the lower on the display this point will be.
3	If cursors are added to the oscilloscope screen and aligned with a corresponding point on the waveform, one complete rotation of the camshaft can be seen. (This is equivalent of two complete revolutions of the crankshaft).

Automotive Sensors & Waveform Analysis

> If a diagnostic trouble code is generated which suggests an implausible signal from the camshaft sensor, or a correlation issue between the camshaft and crankshaft, this may be caused by incorrect valve timing.
> The tension of the camshaft drive belts or chains should be checked, and it should be ensured that their timing marks are still aligned.
> Timing pulleys should be checked for security and the operation of any variable valve control should be fully investigated.

Fuel pressure sensor

Fuel pressure sensors are used on common rail Diesel engines to help monitor and regulate the pressure in the rail depending on operating conditions. The engine management system needs to vary the pressure at the injectors and this will be pre-programmed into the ECU, but in order to cope with sudden driver and operation demands, a **closed loop** feedback is required which tells the system what is occurring in order to make immediate changes if needed. Depending on the manufacturer, the engine management ECU will use information provided by the sensor to control fuel pressure between approximately 280 bar at idle, up to around 1600 bar at full throttle.

> **Closed loop** – an engine management operating state where sensor readings are used to influence the operation of an actuator or system.

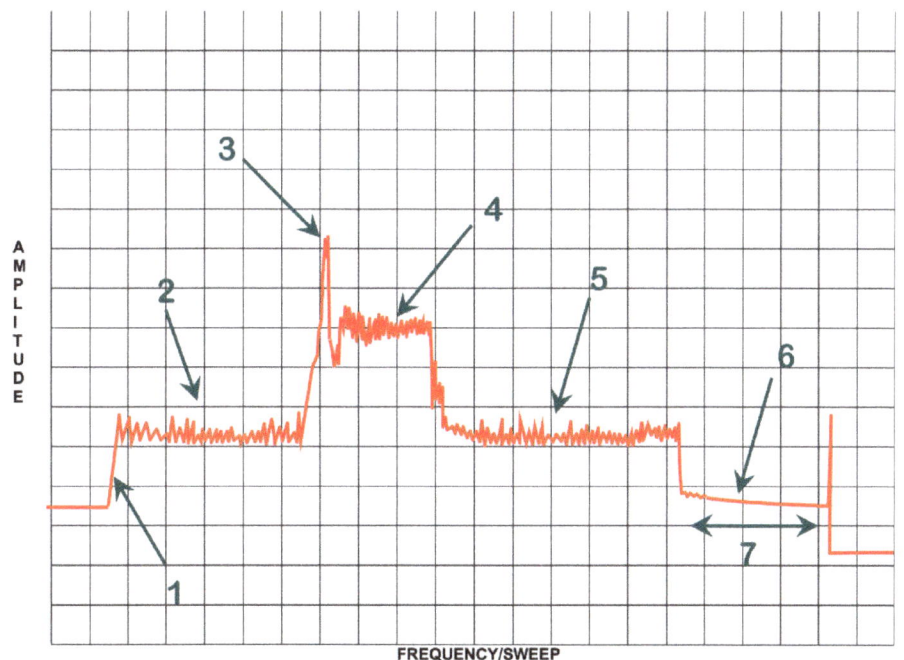

Figure 4.21 Common rail fuel pressure sensor

Automotive Sensors & Waveform Analysis

Table 4.21 Waveform analysis common rail fuel pressure sensor

Waveform component	Description
1	This sudden rise in voltage on the waveform represents a 'key on' condition as voltage is fed to the sensor via the ignition circuit and ECU. At this point, there is no pressure in the fuel rail. This voltage rise is used as a sensor circuit check by the ECU and if 0 volts is registered, the sensor has failed and a diagnostic trouble code may be generated.
2	When the engine is started, and allowed to idle, the voltage will rise to represent the fuel pressure at tick-over.
3	This point on the waveform shows the throttle being snapped open and the fuel pressure being instantly increased to allow for hard acceleration.
4	On an unloaded engine, once the snap throttle opening has brought the engine up to full governed RPM, the fuel pressure settles back to an optimal fast running condition and this is sown by the waveform at this point as approximately half way up the scale, when compared to the maximum amplitude. If the same test was done on a loaded engine, you would expect to see the amplitude of the waveform approximately three-quarters of the way up the scale.
5	When the throttle pedal is released and the engine returns to idle, the height (amplitude) of the waveform settles out to a similar position as waveform component 2 on this image.
6	When the key is switched off, the voltage on the waveform can be seen to drop off over a short period of time as the pressure gradually leaks away, before the ECU switches off power to the sensor.
7	The time period for this section of the waveform, as the engine is switched off should be around ten seconds, and the voltage should fall gradually. If the voltage falls too rapidly at this point, system pressure is escaping, which may indicate a fault such as a leaking injector or high pressure pump.

Knock sensor

The knock sensor is mounted in the cylinder head or cylinder block, normally between the two centre cylinders. It contains a piezoelectric crystal which, when deformed by a vibration, will create a small electric potential. The signal from this sensor is used by the engine management system to help detect 'pinking' or engine 'ping'.

Pinking occurs when the ignition timing of an engine is too far advanced, and as the combustion flame front spreads out within the cylinder, the pressure wave reaches the piston crown when the engine is at top dead centre TDC. As the connecting rod is vertical, little torque or tuning effort is applied to the crankshaft, power is lost and a distinctive knock known as 'ping' occurs. If the ignition timing can be advanced to a point just before 'pinking' then the engine will run with optimum torque.

If 'pinking' is detected by the knock sensor, the corresponding signal will be picked up at the ECU and the ignition timing will be retarded until no knock occurs; at which point, the timing will be once again advanced. By continuously monitoring the knock sensor, the engine management system is able to keep the ignition timing within the required limits.

The signal from a knock sensor is very quick, so the sweep or timeframe of the oscilloscope display will need to be adjusted accordingly in order to pick-up an acceptable waveform.

Automotive Sensors & Waveform Analysis

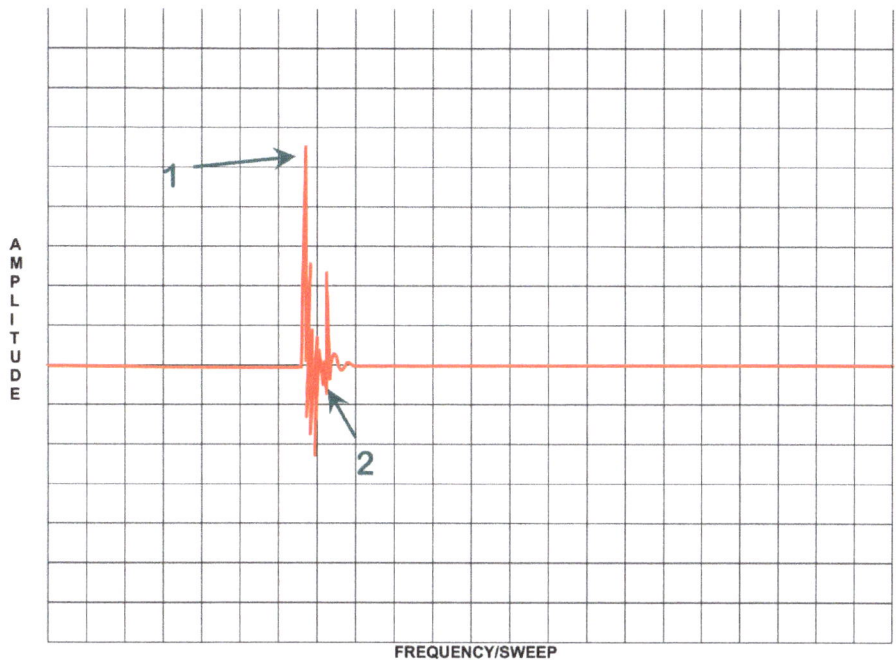

Figure 4.22 Knock sensor

Table 4.22 Waveform analysis knock sensor

Waveform component	Description
1	This is the point on the waveform where a voltage potential is created in the piezoelectric crystal by a knock from over advanced ignition timing. The greater the shockwave, the higher the initial peak.
2	This shows the aftershock as the engine knock dissipates.

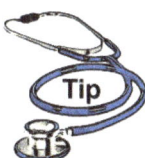

As the knock sensor creates its own voltage potential, it can often be tested without power, and with or without the engine running.
If the sensor is removed, it can be connected to an oscilloscope, and lightly tapped with a metallic tool to create the waveform described.

It is very important to correctly torque the sensor when fitting to the engine, as if it is too tight or too loose, an incorrect signal signature will be produced.

Automotive Sensors
& Waveform Analysis

Another method for checking the function of the knock sensor when the engine is running, is by using an ignition timing light (if a suitable secondary connection can be found).
With the timing light illuminating the crankshaft timing marks, lightly tap the engine block or cylinder head near to the location of the knock sensor with a metallic implement (a spanner often works well for this).
When the shockwave is detected by the ECU, if the sensor is working, the timing will be retarded and this can be seen in the 'shuffling' of the timing marks.

Lambda sensor + heater

The lambda sensor or oxygen sensor, performs a vital role in helping the engine management system maintain the correct air/fuel ratio for effective operation of the catalytic converter.
It is mounted in the exhaust system just in front of the catalytic converter, and compares a sample of the oxygen contained in the exhaust gas to the oxygen contained in a sample of air. If the correct air/fuel ratio is achieved, then nearly all the available oxygen will be used up during the process of combustion.

- If too much oxygen remains in the sample, the engine is running weak.
- If too little oxygen remains in the sample, the engine is running rich.

The engine management system takes this information and 'trims' the open time of the fuel injector to correct the air/fuel ratio.

- Reducing the open time of the fuel injector weakens the mixture.
- Increasing the open time of the injector richens the mixture.

When working correctly, this continuous adjustment of air/fuel ratio from rich to weak, produces a characteristic rising and falling waveform from the lambda sensor.

Titania lambda sensor

A Titania lambda sensor is supplied with voltage from the engine management ECU and when the sensor element comes into contact with the exhaust gasses, it will produce an oscillating waveform with an amplitude of around 0.5 volts to 4 volts as the fuelling is adjusted from lean to rich.
The waveform of a Titania lambda sensor will only begin to oscillate properly when it has reached the correct operating temperature; to assist fast operation from start-up, a heater element is often included in the sensor design.

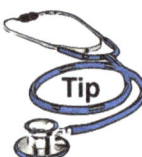

Use a diagnostic scan tool and check readiness monitors to ensure that the engine management systems is running in closed loop and the Lambda sensor is ready to be tested.

Automotive Sensors & Waveform Analysis

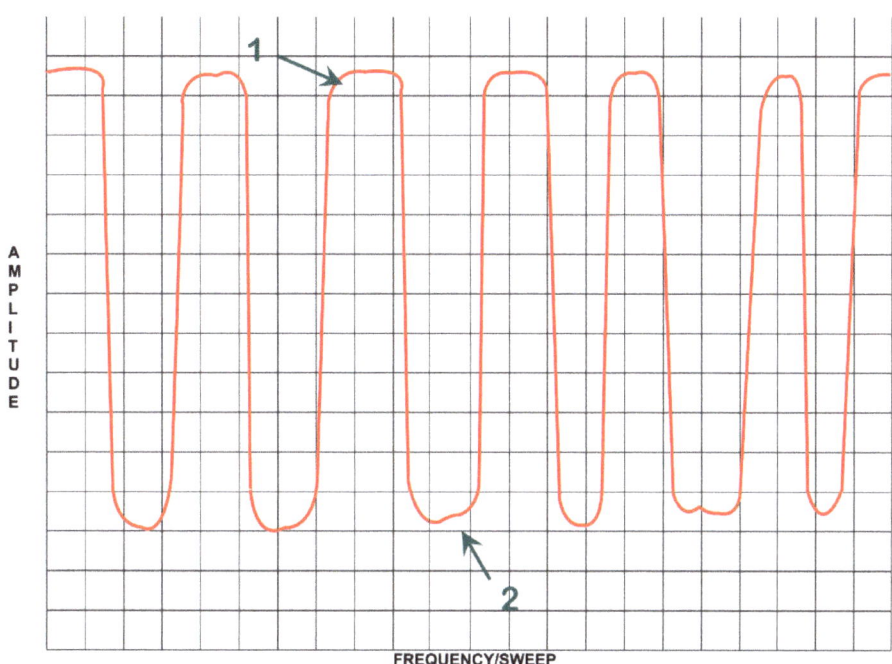

Figure 4.23 Titania lambda sensor

Table 4.23 Waveform analysis Titania lambda sensor

Waveform component	Description
1	This point on the waveform shows the air/fuel ratio as rich and will have an amplitude of approximately 4 volts.
2	This point on the waveform shows a lean air/fuel ratio and will have an amplitude of approximately 0.5 volts.

Zirconia lambda sensor

Unlike a Titania sensor, a Zirconia type lambda sensor will produce its own voltage proportional to the oxygen content of the exhaust gasses, and this information is used by the engine management system to correct air/fuel ratio.

Like the Titania sensor, a Zirconia lambda sensor will only function correctly when it has reached its proper operating temperature and will often have a heater element incorporated in its design to assist fast operation from start-up.

Because the Zirconia sensor produces its own voltage, its overall amplitude will be smaller than that of an equivalent Titania lambda sensor; approximately 0.2 volts when lean and 0.8 volts when rich.

Automotive Sensors & Waveform Analysis

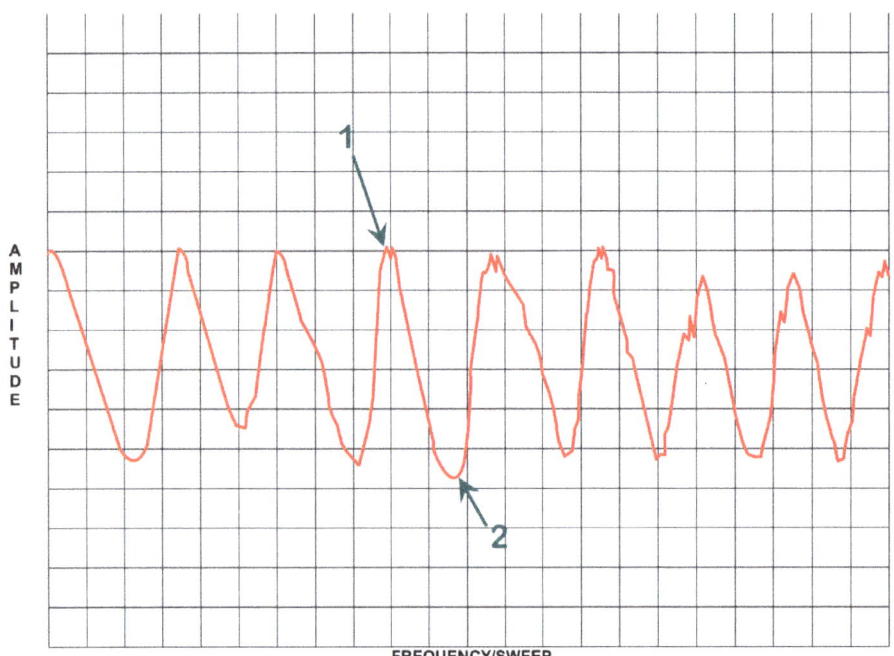

Figure 4.24 Zirconia lambda sensor

Table 4.24 Waveform analysis Zirconia lambda sensor

Waveform component	Description
1	This point on the waveform shows the air/fuel ratio as rich and will have an amplitude of approximately 0.8 volts.
2	This point on the waveform shows a lean air/fuel ratio and will have an amplitude of approximately 0.2 volts.

Tip: The waveform of a correctly functioning lambda sensor, which has reached its proper operating temperature, should oscillate with a frequency of around 1 Hertz; from rich to lean in around one second. (It may be necessary to raise engine speed to around 2000 RPM to produce an adequate oscillation from the sensor).
Over a period of time, a lambda sensor may become dirty, giving a slow response time. It is sometimes possible to remove a lambda sensor and clean with a solvent to improve its operation.

Lambda sensor heater

For a lambda sensor to operate correctly, and give accurate readings to the engine management ECU, it needs to be running at an optimum temperature.
This means that from cold-start and during warm-up the engine management system will ignore any readings from the sensor and operate in an '**open loop**' mode, where pre-programmed parameters are used for fuel injection and air/fuel ratio.
Once the lambda sensor has reached an appropriate operating temperature, the ECU will switch to '**closed loop**' mode and adjust the fuelling in accordance with the data provided.

Automotive Sensors & Waveform Analysis

This means that the quicker a lambda sensor can be brought up to temperature, the sooner emissions will be correctly regulated.

To speed up this process, most lambda sensors are manufactured with an electrically controlled heater element.

Open loop – an operating state when readings from a sensor are ignored as it has not yet reached its required operating condition.

Closed loop – an operating state where sensor readings are used to influence the operation of an actuator or system.

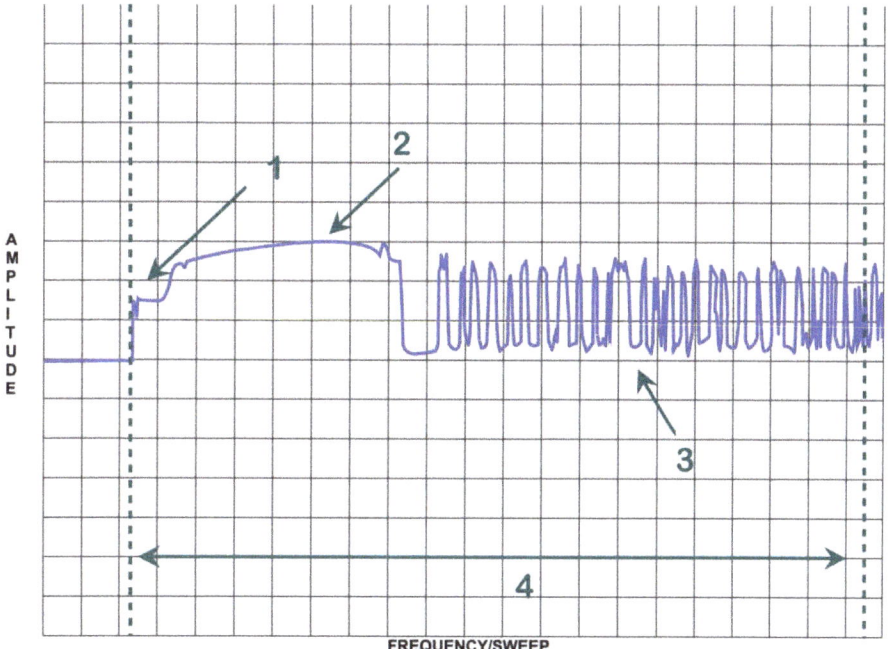

Lambda sensor heater

Connect to the circuit using:

Inductive Amps Clamp

Figure 4.25 Lambda sensor heater current

Table 4.25 Waveform analysis lambda sensor heater

Waveform component	Description
1	With the oscilloscope connected using a current clamp and set-up to measure amps; this shows the point where the ignition is switched on and current quickly flows into the lambda sensor heater circuit.
2	This shows a period of current flow that is used to rapidly bring the heater element up to temperature.
3	To prevent overheating or burning out of the heater element, the current is regulated by pulse width modulation (quickly switching on and off). During this period, the waveform can be seen to rapidly rise and fall.
4	Once up to operating temperature, the heater circuit of the lambda sensor can be switched off. The warm-up period of the lambda sensor is relatively brief, and the timeframe between the cursors represents approximately 20 seconds from start-up.

Automotive Sensors & Waveform Analysis

A faulty heater circuit within a lambda sensor may give no apparent engine operating symptoms, as once it has reached its normal operating temperature, it will provide readings to the engine management ECU. Due to the possibility of this creating an emission related issue, a diagnostic trouble code will be generated, and an engine management malfunction indicator lamp will be illuminated.
If the heater circuit contains its own fuse, current flow can be checked at the fuse box when the ignition is switched on to confirm operation, and this reduces the need to access the lambda sensor itself.
It will also help ensure that the fuse is not blown!

Post-catalytic converter lambda sensors

Many manufacturers include post-catalytic converter lambda sensors in their vehicle designs. The purpose of these post-catalyst sensors is to monitor the physical operation of the converter.
By comparing the signals produced between the two sensors (before and after the catalyst) the engine management system can calculate the efficiency of the converter itself.
During the chemical conversion inside a correctly functioning catalyst, much of the oxygen contained in the exhaust gas is used to change carbon monoxide (CO) into carbon dioxide (CO_2). This means that although the waveform produced by the post-catalytic converter sensor will oscillate, following the pattern of the pre-catalyst sensor, its amplitude will be greatly reduced.
The sensors may be either Titania or Zirconia type.

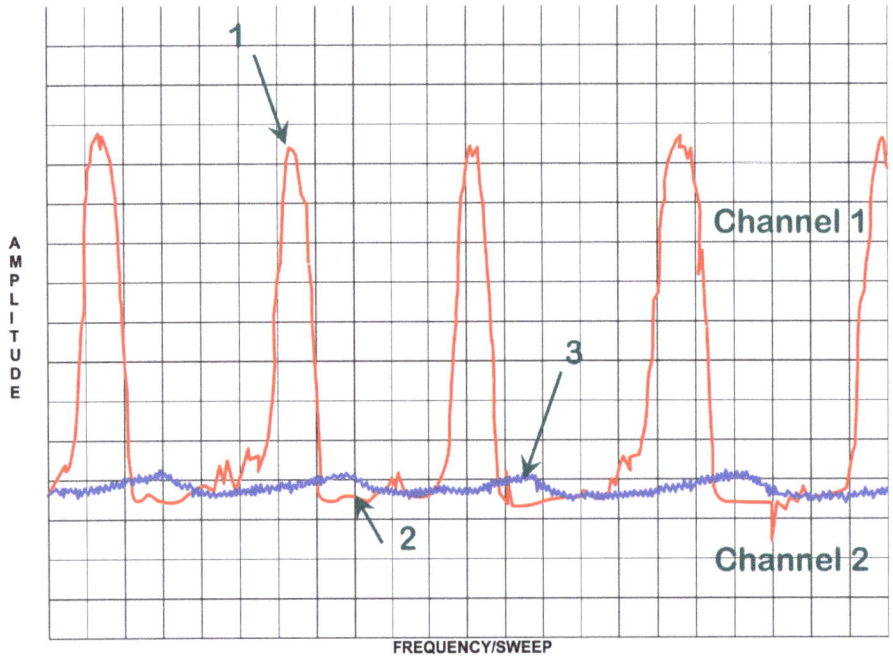
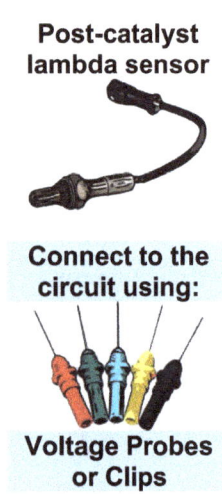

Figure 4.26 Pre and Post catalytic converter lambda sensors

Automotive Sensors & Waveform Analysis

Table 4.26 Waveform analysis Pre and Post catalytic converter lambda sensors

Waveform component	Description
1	Channel 1. This point on the waveform shows the pre-catalyst lambda sensor with the air/fuel ratio rich, and will have an amplitude of approximately 4 volts (Titania) or 0.8 volts (Zirconia).
2	Channel 1. This point on the waveform shows the pre-catalyst lambda sensor with a lean air/fuel ratio, and will have an amplitude of approximately 0.5 volts (Titania) or 0.2 volts (Zirconia).
3	Channel 2. With a second channel connected to the signal wire of the post catalytic converter, the waveform produced will oscillate slightly, following the pattern of the pre-catalyst sensor, however, the amplitude should be much lower (almost flat).

Wideband oxygen sensor

To improve the efficiency of the emission control system, some manufacturers use wideband oxygen sensors. The purpose of a wideband oxygen sensor is the same as any other lambda sensor, to measure the oxygen content of the exhaust gas and relay this information to the engine management ECU, so that the air/fuel ratio can be controlled.

Its construction and operation, however, are slightly different from those described for Titania and Zirconia types. The design of a wideband lambda sensor does not rely solely on the change of resistance in a Zirconia or Titania element; instead, two chambers are created: one containing exhaust gas, and the other open to air for reference. A component called an 'oxygen pump' is embedded in the wideband sensor, and through an electrochemical process, it tries to maintain a stable oxygen quantity in one chamber. The quantity of oxygen in this chamber is measured by a Zirconia ceramic. The amount of current supplied to the oxygen pump to maintain the correct oxygen content in the chamber, is proportional to the amount of oxygen in the exhaust gas.

From this information, the engine management ECU can maintain a very stable air/fuel ratio.

A wideband oxygen sensor can often be identified by the number of wires connecting it to the engine management circuit (five or more).
It is important however, to correctly confirm the type of lambda sensor being used by the manufacturer, before making any diagnostic decisions based on the waveforms produced.

Stoichiometric – a balanced chemical reaction often used to describe the ideal air/fuel ratio.

Automotive Sensors & Waveform Analysis

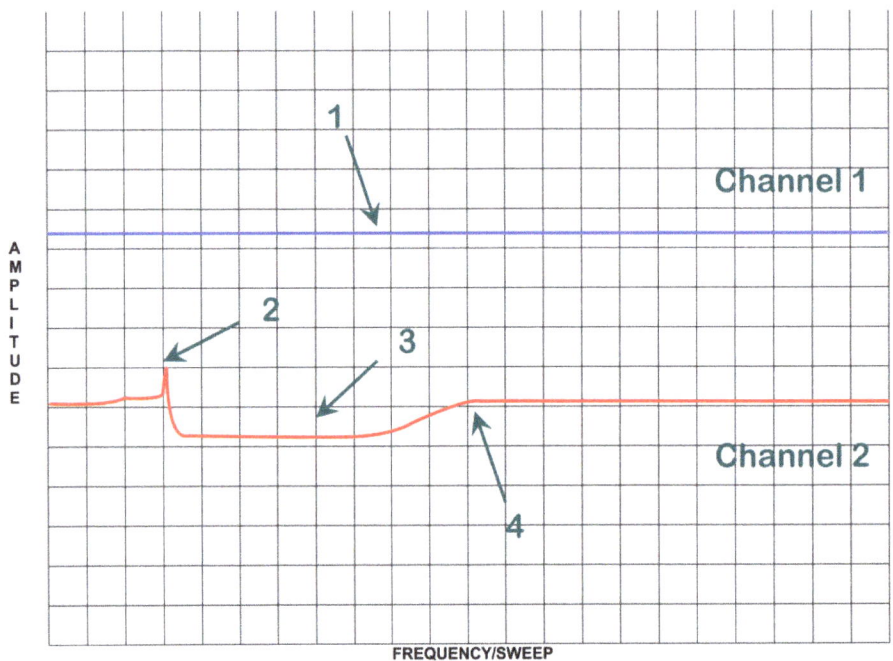

Figure 4.27 Wideband oxygen sensor

Table 4.27 Waveform analysis wideband oxygen sensor

Waveform component	Description
1	Channel 1. This channel on the oscilloscope is connected to the measurement cell of the lambda sensor, and with the engine running, it's voltage should remain relatively stable regardless of engine air/fuel ratio.
2	Channel 2. This channel is connected to the oxygen pump of the wideband sensor, and when running at the correct air/fuel ratio, the voltage will remain stable at 0 volts. When the throttle is snapped open, there is a small spike in voltage as the oxygen content of the exhaust gas falls due to acceleration enrichment.
3	Channel 2. When the throttle pedal is released, the voltage will fall rapidly into a negative value as the oxygen content of the exhaust gas increases and oxygen is pumped out of the measurement cell. The timescale of the oscilloscope may have to be set relatively fast to ensure that the waveform is captured due to the very quick reaction time of the sensor.
4	Channel 2. As the engine management system calculates and resets the fuel injection, the air/fuel ratio returns to its **stoichiometric** value and the voltage returns to zero.

Automotive Sensors & Waveform Analysis

Vehicle speed sensor VSS

The VSS sensor may be used by several vehicle systems to provide information on transmission output or road speed. This data can be shared on a network for engine management, vehicle stability, advanced braking and dashboard information for example.

For accurate measurement, many manufacturers use a Hall effect sensor which produces a digital square waveform.

This sensor is often located at the output of the gearbox, but may sometimes be found at the dashboard speedometer head.

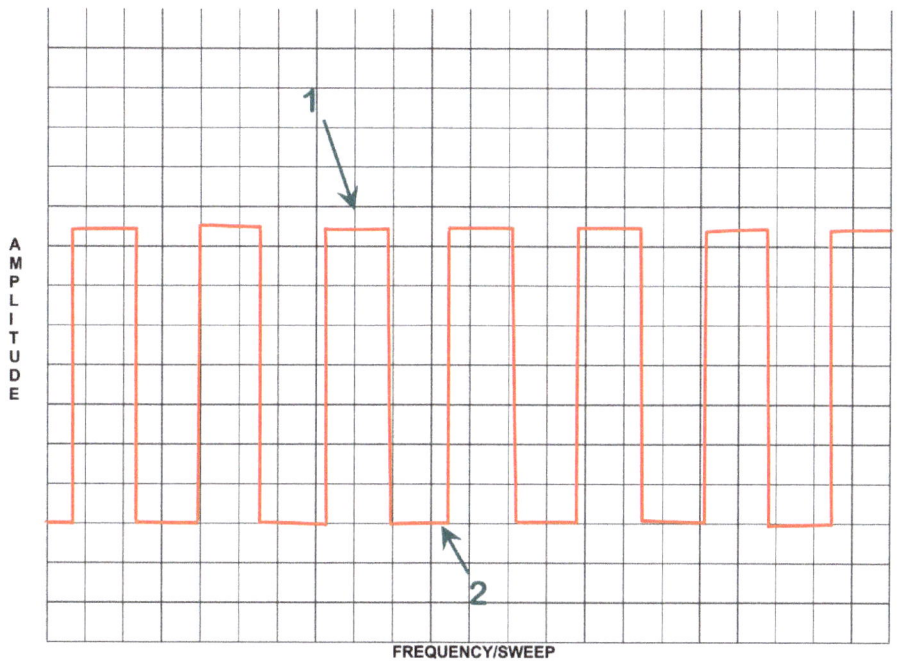

Figure 4.28 Vehicle speed sensor

Table 4.28 Waveform analysis vehicle speed sensor VSS

Waveform component	Description
1	To obtain a waveform, the vehicle will need to be driven or the wheels raised from the floor and allowed to spin freely in drive. This point on the waveform indicates the signal on position and will not change in amplitude (height) with a variation of speed.
2	This point on the waveform indicates the signal off position. As road speed increases, so does the frequency of the digital waveform.

Automotive Sensors & Waveform Analysis

Sun load sensors

A sun load sensor is used to help maintain occupant cabin temperature in conjunction with a climate control system. The sun load sensor is a small photoelectric cell. It is normally mounted on the dashboard of the car behind the windscreen.

In direct sunlight, you will feel hotter than the actual ambient cabin temperature due to the radiation produced by the sun's rays. It is the job of the sun load sensor to pick-up this added solar radiation and adjust the loading on the climate control system to compensate, regardless of passenger compartment temperature.

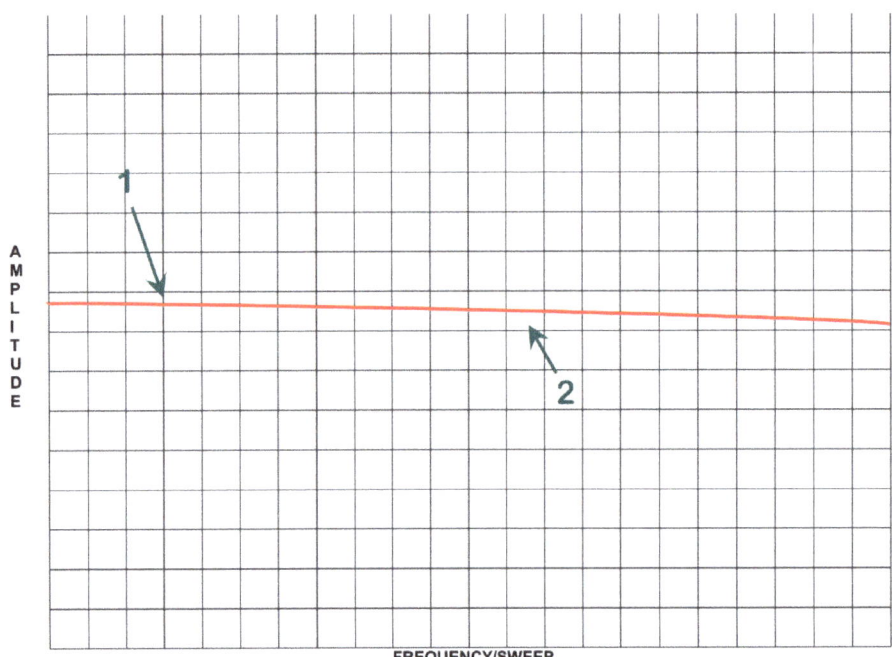

Figure 4.29 Sun load sensor

Table 4.29 Waveform analysis sun load sensor

Waveform component	Description
1	This point on the waveform shows a low sun load on the sensor, indicating less solar radiation and could be simulated by shading the sensor from direct sunlight.
2	This point on the waveform shows increased sun load on the sensor, indicating more solar radiation and could be simulated by placing in direct sunlight.

Tip: Be careful when testing and diagnosing sun load sensors; they do not react to artificial light, such as fluorescent light.
If you try to simulate an increase or decrease in radiation using fluorescent light to test the sensor indoors, it may show very high resistance or open circuit.

Automotive Sensors & Waveform Analysis

Steering angle and torque sensor

The steering angle and torque sensor is used by a number of vehicle systems to determine the direction and speed that the driver turns the steering wheel. This information is then shared on the vehicle network so that actions can be taken by the chassis system to control power assisted steering and vehicle stability etc.
It often works by having two sensors mounted on the steering column, separated by a torsion bar.
The most common type of steering angle sensor (SAS), uses an LED light source, a pick-up and an interrupter disc. As the steering is turned from side to side, a digital signal is created by the pick-up as the light source is received or interrupted.
By comparing the frequency of the digital signal and the alignment of the waveforms created by the upper and lower sensors, the ECU can calculate the speed and direction of rotation from the steering wheel. Also by measuring how far out of **phase** the two signals are, the ECU will also calculate the turing effort (torque) input from the driver.

Phase – the alignment of two or more waveforms, as displayed on an oscilloscope screen for example. ('in phase' the waveforms are lined up; 'out of phase' the waveforms are misaligned).

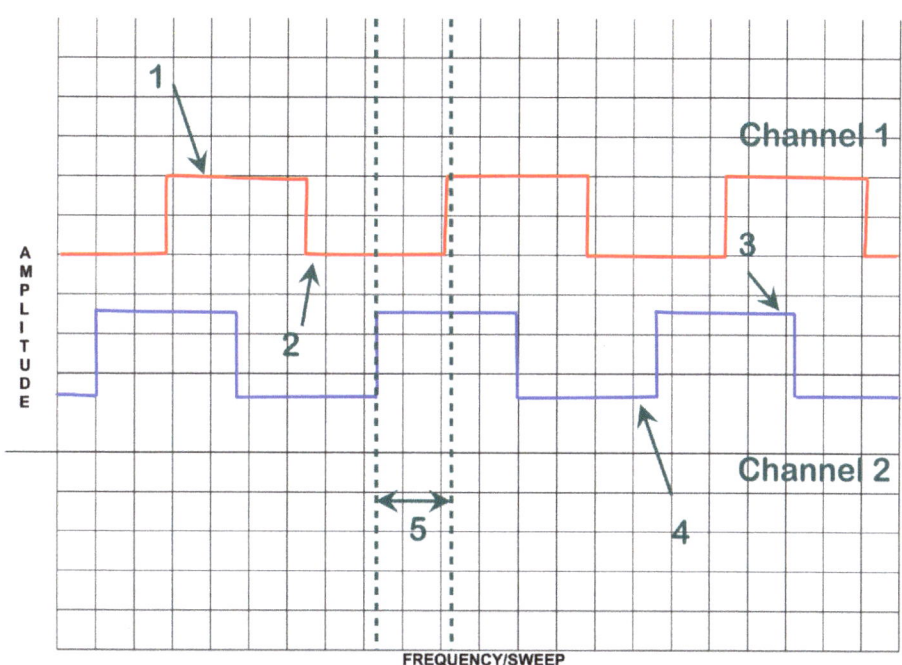
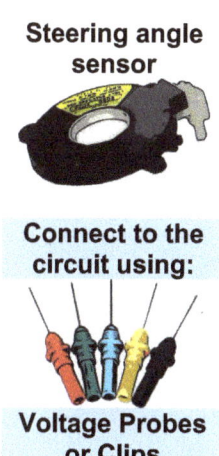

Figure 4.30 Digital steering angle sensor

Automotive Sensors & Waveform Analysis

Table 4.30 Waveform analysis digital steering angle sensor

Waveform component	Description
1	Channel 1. This is connected to the upper steering column sensor, and shows light from the LED being picked up by the photosensitive receiver.
2	Channel 1. This point on the waveform shows light from the LED being blocked by the interrupter disc.
3	Channel 2. This is connected to the lower steering column sensor and shows light from the LED being picked up by the photosensitive receiver.
4	Channel 2. This point on the waveform shows light from the LED being blocked by the interrupter disc.
5	Channels 1 and 2. The difference between the two cursors displayed on the oscilloscope screen shows that the waveforms are 'out of phase' with each other. The direction of phase shift shows which way the steering is being turned, and the amount of phase shift gives an indication of how much torque or turning effort is being applied by the driver.

Automotive Ignition Systems
& Waveform Analysis

Chapter 5 Automotive Ignition Systems and Waveform Analysis

This chapter will help you develop knowledge and understanding of automotive petrol ignition systems. It will enable you to conduct effective diagnosis and repairs of system faults; supporting you by providing a breakdown of waveform images and analysing why the patterns are formed. Remember to work in a systematic way, and observe the relevant environmental, health and safety regulations at all times.

Contents

Introduction	133
Ignition primary waveform	135
Ignition primary current	136
Primary voltage verses current	138
Ignition secondary	141
DIS Secondary negative fired	143
DIS Amplifier earth	145
3 wire coil on plug	146
Coil on plug secondary verses digital trigger	148

There are many hazards associated with the service and maintenance of light vehicle electrical and electronic systems. You should always assess the risks involved with any diagnostic, maintenance or repair routine before you begin and put safety measures in place.
You need to give special consideration to the possibility of:
• The risk of electric shock, especially when working on or around secondary ignition voltages.
• The hazards associated with running engines in confined spaces.
You should always use appropriate personal protective equipment (PPE) when you work on these systems. Make sure that your selection of PPE will help protect you from these hazards.

Don't forget your PPE and VPE

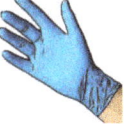

Automotive Ignition Systems & Waveform Analysis

Information sources

The complex nature of light vehicle electric and electronic systems requires a good source of technical information and data. In order to conduct diagnostic, maintenance and repair procedures, you need to gather as much information as possible before you start.

Sources of information may include:

Table 5.1 Possible information sources

Verbal information from the driver	Vehicle identification numbers
Service and repair history	Warranty information
Vehicle handbook	Technical data manuals
Workshop manuals/Wiring diagrams	Safety recall sheets
Manufacturer specific information	Information bulletins
Technical helplines	Advice from other technicians/colleagues
Internet	Parts suppliers/catalogues
Jobcards	Diagnostic trouble codes
Oscilloscope waveforms	On vehicle warning labels/stickers
On vehicle displays	Reference/Textbooks

Always compare the results of any inspection, testing or diagnosis to suitable sources of data. Remember that no matter which information or data source you use, it is important to evaluate how useful and reliable it will be to your safety, diagnostic, maintenance and repair routine.

Where to start?

Step 1
- A good systematic diagnostic routine should always begin with careful questioning of the driver to gather as much information about the symptoms and history of the fault.

Step 2
- After a brief visual inspection for obvious signs of damage or safety issues, the system should be tested to try and recreate the fault.

Step 3
- The vehicle should then be scanned for diagnostic trouble codes and any codes should be recorded. (If possible, a full scan should be conducted, as issues in unrelated systems can sometimes affect the operation of others).

Step 4
- Any codes should be cleared and the vehicle should be tested over a complete drive cycle.

Step 5
- Rescan the vehicle and concentrate diagnosis around any codes that have returned.

Step 6
- Connect the oscilloscope to the suspected circuit and analyse any waveforms produced.

Automotive Ignition Systems & Waveform Analysis

Ignition system

The vehicle ignition system of a petrol-powered engine is designed to create the high voltages needed to produce a spark at the spark plug.

Extra care will be required when working on or around vehicle ignition systems as the high voltages produced may cause personal injury or equipment damage. Ensure that you follow all health and safety guidelines relating to high voltage testing, including the connection of any required grounding test leads on the oscilloscope equipment.

In order to safely and correctly measure ignition system waveforms, you may need to attach special adapters to your oscilloscope.
During the generation of sparks at the plugs, very high voltages are created within the electrical wiring of the circuits.
- An attenuator is an adapter which is connected in series with the test leads of an oscilloscope. It's a small resistor which reduces the height (amplitude) of the waveform by a set amount and allows it to displayed correctly on the screen. Although this device will allow the full height of the waveform to be shown on the screen, you will have convert the graduations on the Y axis to allow you to calculate the correct voltage.
- The voltages flowing through the secondary circuit to the spark plugs are too high to be measured using conventional test leads on the oscilloscope. Most manufacturers of automotive oscilloscopes provide inductive test leads that clamp around or are held against the secondary circuit and can pick up on the high voltage pulses and display them as a waveform on the oscilloscope screen.

Vehicle Ignition Systems

The purpose of an ignition system is to create a spark in the cylinder of a petrol engine, to initiate the combustion of the fuel. This is achieved by taking low battery voltage of around 12 volts and stepping-it-up to many thousands of volts in an ignition coil. It is then discharged across the electrode gap of a spark plug to ignite the air/fuel mixture.

Distributors

Early petrol engine ignition systems used a distributor to time and allocate the sparks at the plugs. It consisted of a switch method to charge and discharge the ignition coil, and a rotor arm which would distribute the high-tension voltage to the correct spark plug lead in the firing order of the engine.
The trigger method employed to switch the ignition coil on early models used a contact breaker (points) as a mechanical system of control. Later distributors used an electronic pick-up (inductive or Hall effect) to signal an ignition module to trigger the ignition coil.

Automotive Ignition Systems & Waveform Analysis

DIS/Wasted Spark

A distributorless (DIS) ignition system is one, as the name suggests, that doesn't have a distributor. Instead it uses the wasted spark principle to discharge the ignition coil through two companion cylinders at the same time. If an engine uses a '**flat-plane crankshaft**', two pistons will move up and down together. If the spark is discharged as these companion pistons both approach top dead centre, the one on the compression stroke will ignite the fuel and initiate combustion, while the one on the exhaust stroke will be wasted.

Once the crankshaft has turned through one complete rotation, the process is repeated on the opposite cylinders. This simplifies the timing and allocation of the sparks considerably.

In early wasted spark systems, the companion plugs were attached to either end of the secondary winding and used the metal of the cylinder head to complete the circuit. This meant that each pair of plugs had its own individual coil (often packaged in one casing) and that one plug received a positive spark, while the other received a negative spark.

Flat plane crankshaft – a crankshaft layout where the crank pins are arranged 180° apart.

Coil-On-Plug COP

Many modern ignition systems use a coil-on-plug COP arrangement. This is where each spark plug has its own individual ignition coil mounted directly on top. They can be separate or arranged as a cassette that joins all the coils together. They are often wired using the wasted spark principle so that they fire in pairs on companion cylinders, however, unlike early systems they will each receive a positive spark.

Primary circuit

The primary circuit of an ignition system carries the **low-tension LT** voltage from the vehicles battery. When switched on, the current flow generates an invisible magnetic field which builds up around the windings inside the ignition coil. When the circuit is switched off, the magnetic field rapidly collapses, causing a high voltage of several hundred volts to be induced in the primary winding.

Secondary circuit

The secondary circuit of an ignition system carries the **high-tension HT** voltage to the engines spark plugs. When the primary circuit of the coil is switched off, the collapse in the magnetic field surrounding the primary and secondary coils, will induce a very high voltage of several thousand volts in the secondary winding, and discharge this across the gap of the spark plug.

Transformers

A transformer is an electronic component designed to increase or decrease the amount of electric voltage in a circuit. An oscillating electric current in a primary circuit is used to induce a voltage in a secondary winding. Depending on the size of the secondary winding when compared to the primary, it will either step-up or step-down the voltage for use in an electric circuit.

Automotive Ignition Systems & Waveform Analysis

Low-tension LT – a term used to describe the low voltage circuit of a vehicle ignition system.

High-tension HT – a term used to describe the high voltage circuit of a vehicle ignition system.

Ignition primary waveform

The purpose of an ignition coil is to take the 12 volts in a vehicle battery and step it up to the many thousand volts required to create an effective spark at the spark plug. It does this by generating a magnetic field around two coils of copper wiring and then allowing those magnetic fields to rapidly collapse, which will then induce a much higher voltage within the coils.

With a good understanding of the operation of an ignition coil, a great deal of diagnostic information can be gained by studying the patterns shown from an oscilloscope waveform.

The primary coil is the first part of the process, and carries the 12 volts from the vehicle battery, creating the initial magnetic field when switched on, and current is allowed to flow.

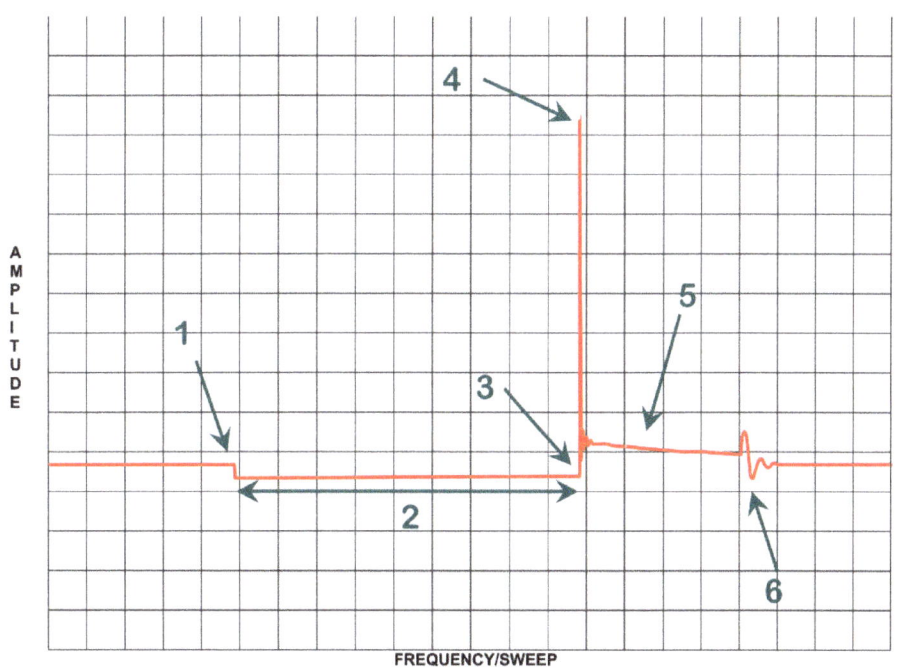
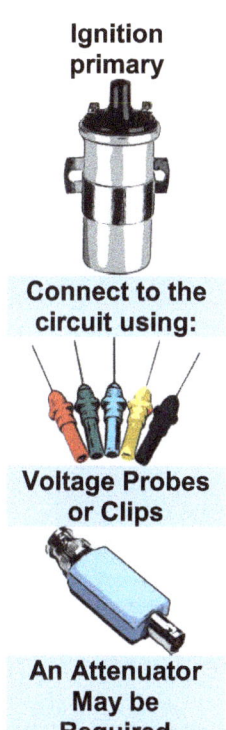

Figure 5.2 Ignition primary coil waveform

Automotive Ignition Systems & Waveform Analysis

Table 5.2 Waveform analysis ignition primary coil

Waveform component	Description
1	The point shown on the waveform here is the ignition coil being switched-on by connecting the primary circuit to earth. This is achieved using an electronic control unit, or in a conventional ignition system, by closing a set of contact breakers (points). Current flows through the primary circuit and a magnetic field is generated.
2	This shows the charge time of the ignition coil, where the magnetic field builds up to a strength suitable for creating a spark when switched off. This charge time is often called the '**dwell period**', originally referring to the time that the contact breakers remained closed or were 'at rest'.
3	At this point the current flow through the primary winding is switched-off by breaking the circuit.
4	When the primary circuit is switched off, either by the electronic control unit or by opening the contact breakers, current flow stops immediately and the rapid collapse in the magnetic field surrounding the coil induces a high voltage in the primary winding. This induced voltage is normally in excess of 300 volts. A high voltage peak at this point can give an indication of a strong magnetic field and good conversion into induced electricity. If the vehicle has more than one coil in its ignition layout, by using multiple channels, the signals can be compared and weak or strong coils may be identified.
5	The sloping trace at this point on the waveform shows the spark line and can be measured using cursors to give a close approximation of the burn time.
6	This point shows the coil oscillations as the induced voltage dissipates (like the bouncing of an un-damped suspension spring) before settling out at battery voltage. The amount of oscillations here gives a good guide to the condition of the ignition coil and should show 4 to 5 peaks (upper and lower). If there are less oscillations than expected, this can give you an indication that the coil should be replaced.

Dwell period – the charge or 'switched on' time of the ignition coil primary circuit.

Ignition primary current

If an inductive current clamp is placed around the primary winding low tension wire, a waveform can be obtained which corresponds to the flow of electricity through the coil, helping to illustrate what is happening during the dwell period of the ignition cycle. Two styles of waveform could be obtained depending on the manufacturer design; current limited and non-current limited.

Automotive Ignition Systems & Waveform Analysis

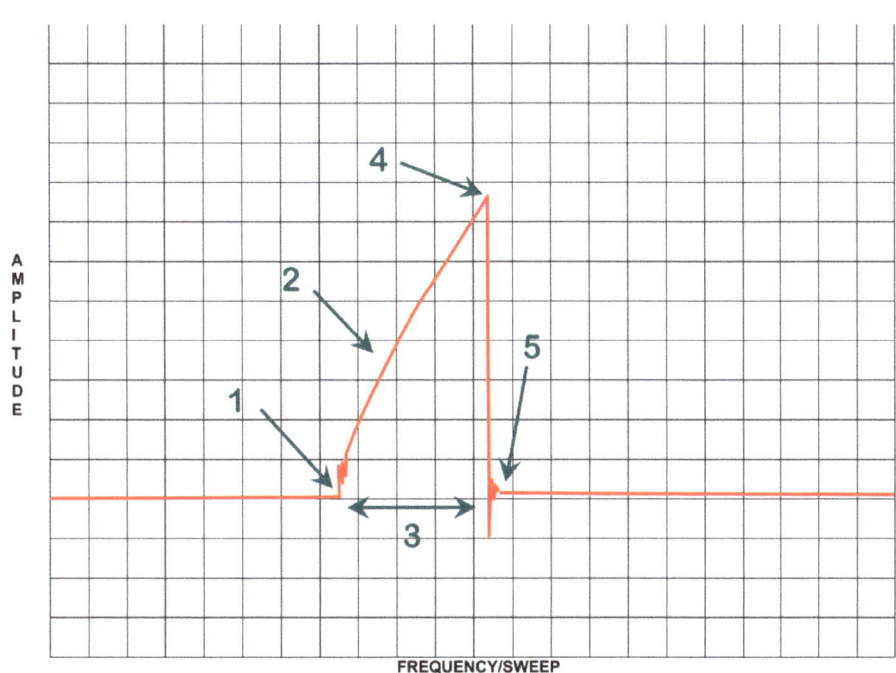

Figure 5.3 Non-current limited primary waveform

Table 5.3 Waveform analysis non-current limited primary coil

Waveform component	Description
1	With a current clamp mounted around the low-tension wire to the ignition coil, this shows the point at which the coil is switched on and current starts to flow.
2	In a non-current limited ignition coil, the flow of current should steadily rise in a relatively smooth curve. If the rise in current is too quick at this point, it gives an indication that the resistance of the coil is too low, leading to overheating and breakdown. If the rise in current is too slow at this point, it gives an indication that the resistance of the coil is too high, leading to poor performance, a weak spark, or misfire.
3	The measurement between these two points shows the dwell period or charge time of the ignition coil. The dwell period will expand with engine RPM.
4	This is the maximum current flow at the time the coil is switched off.
5	This shows the current flow has stopped and the spark has occurred.

Automotive Ignition Systems & Waveform Analysis

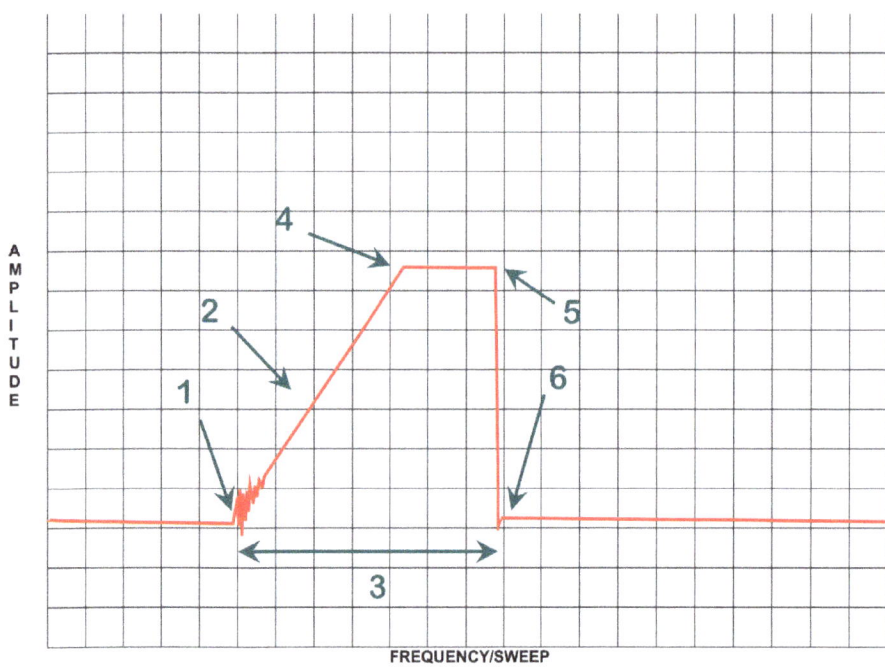

Figure 5.4 Current limited primary waveform

Table 5.4 Waveform analysis current limited primary coil

Waveform component	Description
1	With a current clamp mounted around the low-tension wire to the ignition coil, this shows the point at which the coil is switched on and current starts to flow.
2	In a current limited ignition coil, the flow of current should quickly rise in a relatively smooth curve. In this type of ignition, the internal resistance has been reduced in order to allow the coil to charge very quickly, and this period will remain constant regardless of engine speed.
3	The measurement between these two points shows the dwell period or charge time of the ignition coil. The dwell period will expand with engine RPM.
4	Due to the low internal resistance and the quick charge time of the coil, if left unchecked, the winding would overheat and burn out. As soon as the coil has reached full saturation, the current flow is regulated and held at its maximum by the ECU, making the waveform flatten out at this point.
5	This is the maximum current flow at the time the coil is switched off.
6	This shows the current flow has stopped and the spark has occurred.

Primary voltage verses current

By operating two channels on the oscilloscope, the voltage and the current flow can be compared, providing further diagnostic information.

Automotive Ignition Systems & Waveform Analysis

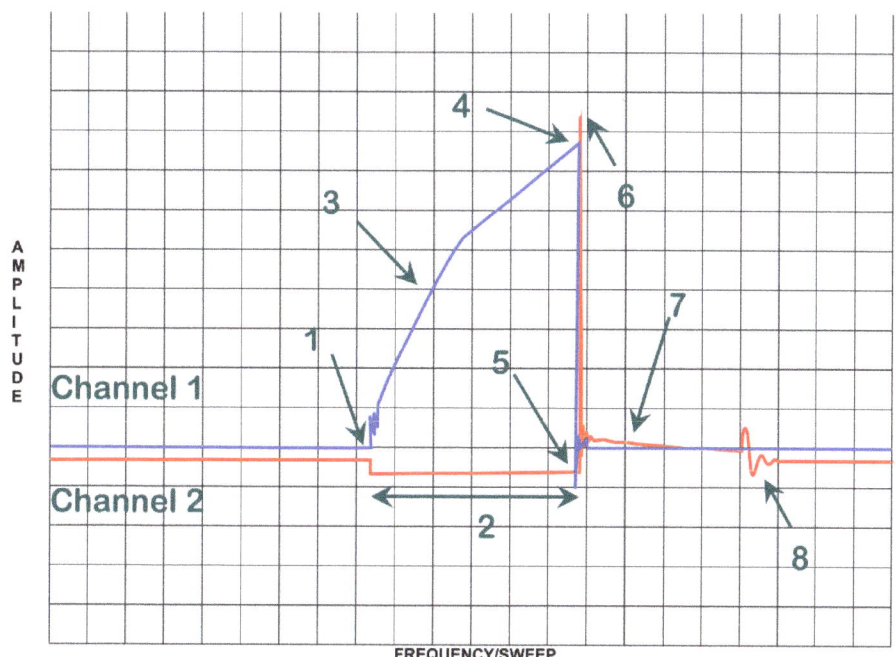
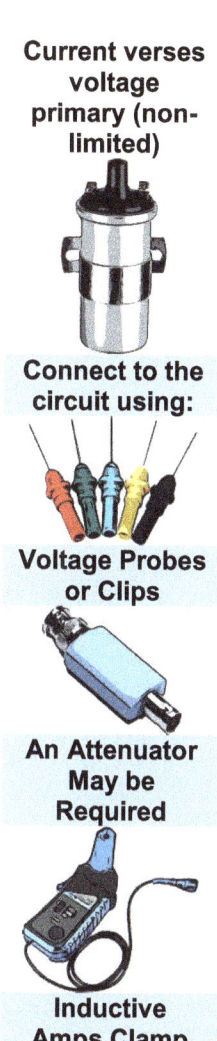

Figure 5.5 Non-current limited verses voltage primary waveform

Table 5.6 Waveform analysis current verses voltage primary (non-limited)

Waveform component	Description
1	This is the point of the waveform, where the coil is switched on and can be seen in both the voltage and current patterns.
2	Channel 1 and 2. This section shows the charge time or 'dwell period' and will vary with engine RPM.
3	Channel 1. In a non-current limited ignition coil, current flow will rise in a relatively steady curve until it reaches its maximum.
4	Channel 1. This shows the maximum current achieved, as the primary circuit in the ignition coil is switched off.

Automotive Ignition Systems & Waveform Analysis

Table 5.6 Waveform analysis current verses voltage primary (non-limited)

5	Channel 1. The current may be drawn momentarily into a negative value as the ignition coil discharges.
6	Channel 2. This shows the peak voltage created in the primary circuit as the ignition coil discharges and creates a spark.
7	Channel 2. This is the spark line or burn time and gives an indication of the duration of the spark.
8	Channel 2. This shows the coil oscillations as the induced voltage dissipates (like the bouncing of an un-damped suspension spring) before settling out at battery voltage. The amount of oscillations here gives a good guide to the condition of the ignition coil and should show 4 to 5 peaks (upper and lower). If there are less oscillations than expected, this can give you an indication that the coil should be replaced.

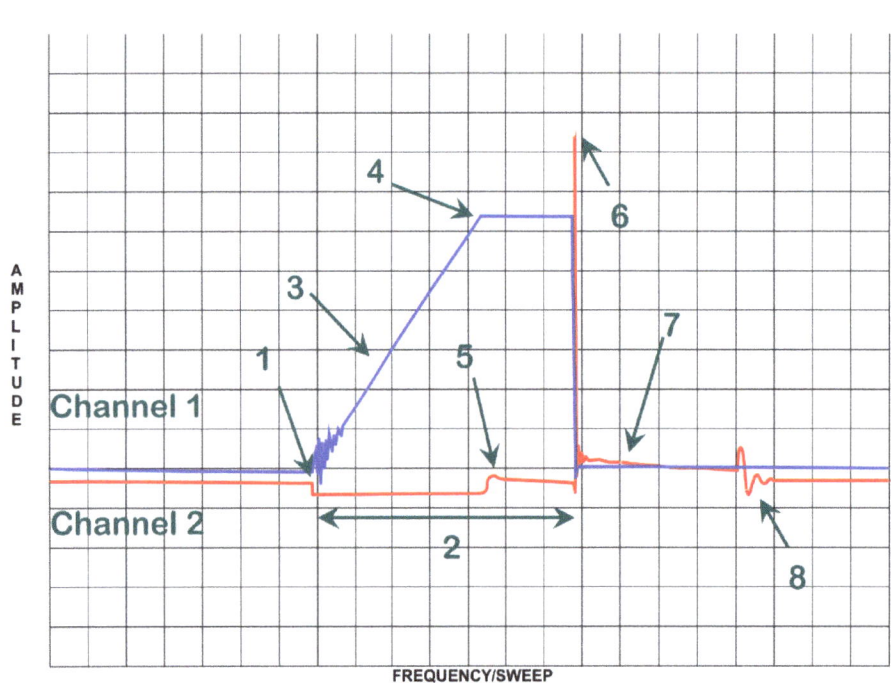

Figure 5.7 Current limited verses voltage primary waveform

Automotive Ignition Systems & Waveform Analysis

Table 5.7 Waveform analysis current limited verses primary waveform

Waveform component	Description
1	Channel 1 and 2. This is the point of the waveform, where the coil is switched on and can be seen in both the voltage and current patterns.
2	Channel 1 and 2. This section shows the charge time or 'dwell period' and will vary with engine RPM.
3	Channel 1. In a current limited ignition coil, the flow of current should quickly rise in a relatively smooth curve. In this type of ignition, the internal resistance has been reduced to allow the coil to charge very quickly, and this period will remain constant regardless of engine speed.
4	Channel 1. Due to the low internal resistance and the quick charge time of the coil, if left unchecked, the winding would overheat and burn out. As soon as the coil has reached full saturation, the current flow is regulated and held at its maximum by the ECU, making the waveform flatten out at this point.
5	Channel 2. The small hump created in the primary voltage waveform is caused by the current limiting within the winding, controlled by the ECU. As the engine speeds up, this will move to the right and may disappear completely from view at very high engine revs.
6	Channel 2. This shows the peak voltage created in the primary circuit as the ignition coil discharges and creates a spark.
7	Channel 2. This is the spark line or burn time and gives an indication of the duration of the spark.
8	Channel 2. This shows the coil oscillations as the induced voltage dissipates (like the bouncing of an un-damped suspension spring) before settling out at battery voltage. The amount of oscillations here gives a good guide to the condition of the ignition coil and should show 4 to 5 peaks (upper and lower). If there are less oscillations than expected, this can give you an indication that the coil should be replaced.

Ignition secondary

In order to test the secondary ignition circuit using an oscilloscope, a special inductive adapter will need to be used. These adapters will vary in design, but are able to pick up the magnetic pulse created by the high-tension voltage. Some adapters clip around the spark plug leads, while others act as a type of 'wand' which is simply held against the coil or lead as required. The inductive adapter may have a ground lead attached, which is used to protect the user and equipment from a high voltage shock, should the spark try and find an easier path to earth (short circuit). The settings on the oscilloscope will need adjusting to a secondary circuit and inductive sensor, which will then give the amplitude in kilovolts.

Automotive Ignition Systems & Waveform Analysis

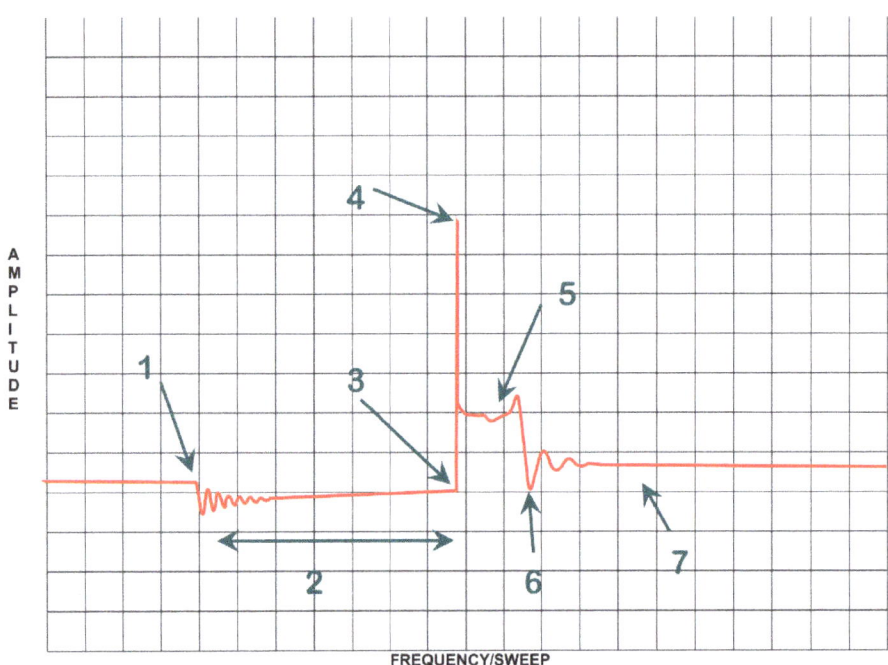

Figure 5.8 Ignition coil secondary waveform

Table 5.8 Waveform analysis ignition secondary circuit

Waveform component	Description
1	This is where the primary circuit switches on and the ignition coil starts to charge.
2	This is the 'dwell period' or charge time of the ignition coil.
3	This point on the waveform shows when the coil is switched off (either by the contact breakers opening in a conventional ignition system, or by an electronic control unit).
4	Once the coil is switched off, the rapid collapse in the magnetic field induces a high voltage in the secondary winding. The voltage rises until it reaches a point where it can overcome the resistance of the air gap at the tip of the spark plug. The maximum amplitude achieved at this point gives an indication of how much voltage is required to fire the spark plug (shown in kilovolts KV) and should be relatively even when comparing different cylinders. If the ignition system is a wasted spark style, the firing voltage may jump up and down, due to the difference in compression pressures when the cylinder is on its exhaust stroke.
5	This section shows the spark line or burn time of the plug. Once the initial resistance of the air gap at the tip of the plug has been overcome, it takes less voltage to maintain the spark, so the amplitude falls and a fairly shallow wave is shown sloping slightly downwards. If the spark line is at too greater angle on one cylinder, when compared with the others, this can indicate a fault with the spark plug or HT lead.

Automotive Ignition Systems & Waveform Analysis

Table 5.8 Waveform analysis ignition secondary circuit

6	When the spark ends, an aftershock of coil oscillations is created by a dissipating magnetic field.
7	The firing of the spark plug is now complete and the waveform settles out to a flat line before the next action.

High firing voltages may be caused by:
- Large plug gaps
- Break in the plug lead
- Worn spark plugs
- Lean mixture
- Fouled spark plugs
- High compression

Low firing voltages may be caused by:
- Small plug gaps
- Low compression
- Rich mixture
- Incorrect timing
- Tracking to earth
- Fouled spark plugs

DIS Secondary negative fired

On some distributorless ignition systems, the principle of 'wasted spark' is used.
In this design, one ignition coil is used to operate a pair of companion cylinders on an engine. (Companion cylinders are those where the pistons reach top dead centre TDC at the same time). As one of the cylinders will be on its compression stroke, the spark provided will ignite the mixture and initiate the power stroke; its companion cylinder will be on the exhaust stroke, meaning this spark will be wasted.
The main reason a wasted spark system is used by some manufacturers is that it reduces the complexity of the ignition circuit and its operation; you don't have to choose which of the companion cylinders needs the spark, you just have to know when. This also reduces the complexity of engine management programming by half.
As both ends of the secondary coil are attached to a spark plug, they form a circuit through the metal of the cylinder head. When fired, voltage is induced in the secondary winding and current flows in the circuit in one direction, causing one plug to operate with a positive spark and the other with a negative spark. This design may cause uneven erosion of the spark plug electrodes, but has no effect on the ignition of the mixture.

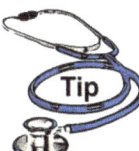

As the two plugs on companion cylinders are used to form a circuit in a DIS wasted spark system, if one plug fails, the circuit is broken, meaning the other plug cannot fire.
If a misfire is detected in a pair of companion cylinders, the circuit and plugs should be inspected to diagnose the fault.

Automotive Ignition Systems & Waveform Analysis

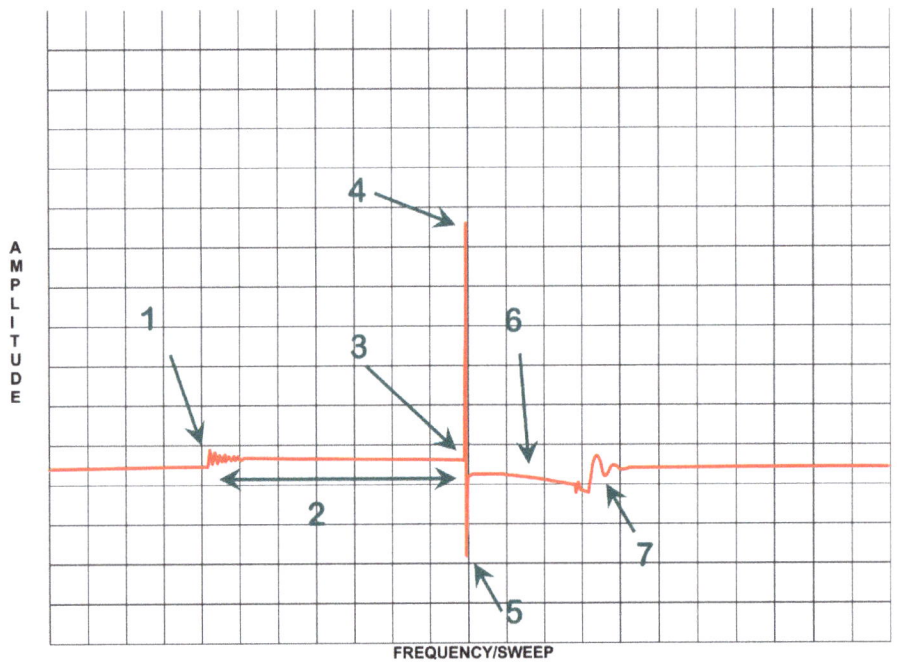

Figure 5.9 DIS negatively fired secondary waveform

Table 5.9 Waveform analysis DIS secondary ignition circuit

Waveform component	Description
1	This is the point on the waveform where the coil switches on and begins to charge.
2	This is the 'dwell period' or charge time of the ignition coil.
3	This point on the waveform shows when the coil is switched off by the electronic control unit.
4	This peak amplitude can sometimes be seen on a negatively fired coil and is the result of the wasted spark process used in this ignition system design.
5	This negative spike is caused by the firing of the plug and is the voltage required to initiate the spark. Its peak negative amplitude will fluctuate because every other stroke, the exhaust valve will be open causing low compression and a wasted spark. The low compression compared with the event stroke does not require as much voltage to initiate the spark.
6	This section shows the spark line or burn time of the plug. Once the initial resistance of the air gap at the tip of the plug has been overcome, it takes less voltage to maintain the spark, so the amplitude rises and a fairly shallow wave is shown sloping slightly up.
7	When the spark ends, an aftershock of coil oscillations is created by a dissipating magnetic field.

Automotive Ignition Systems
& Waveform Analysis

DIS Amplifier earth

Some distributorless ignition systems are fitted with a separate amplifier (also known as a module), which takes a very low voltage trigger signal and uses it to initiate the switching of the coil to create the spark. The ignition coil is normally switched to ground by the amplifier and this is why its earth circuit is so important. In order to correctly test the earth circuit from an amplifier, it must be 'under load' and can only be effectively tested while the engine is running, therefore an oscilloscope is the ideal tool to conduct a diagnosis of this component. The test lead of the scope should be connected to the earth wire of the module and the engine run under various operating conditions.

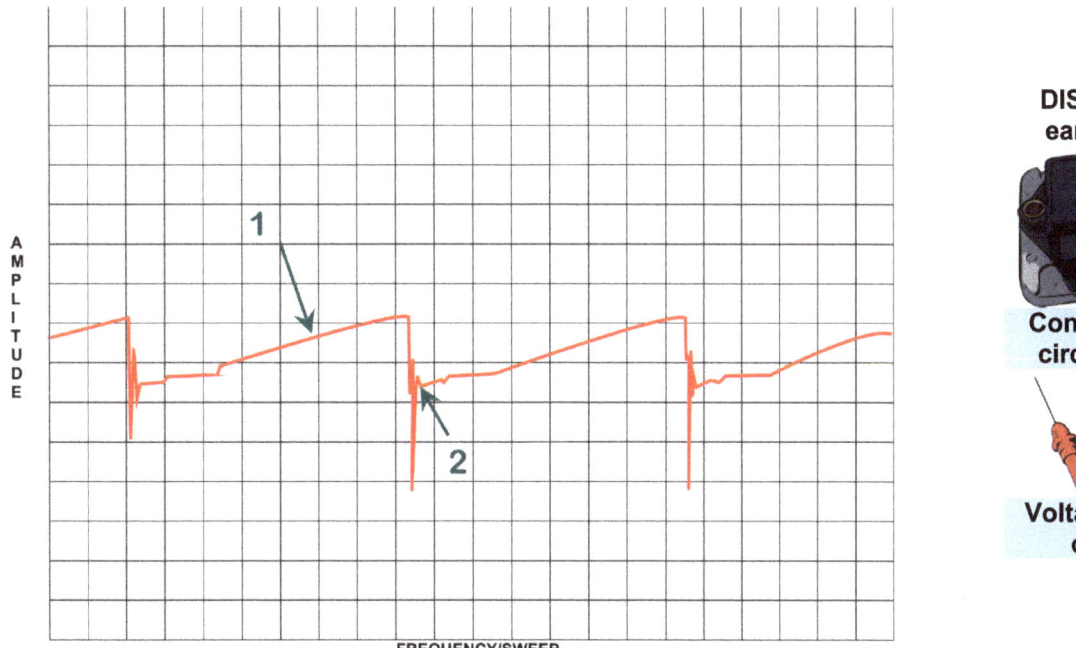

Figure 5.10 DIS amplifier earth circuit waveform

Table 5.10 Waveform analysis DIS amplifier earth circuit

Waveform component	Description
1	This is the important part of the waveform, which shows a small rise in amplitude, representing current in the amplifier earth circuit during the dwell period. In order for the earth circuit to function correctly, there should be no more than around half a volt rise during this period. The flatter the waveform at this point the better. If the voltage rise is too great, the earth circuit should be investigated and repaired.
2	This shows the point on the waveform where the coil is switched off creating the spark.

Automotive Ignition Systems & Waveform Analysis

3 wire coil on plug

When testing a three-wire coil on plug system, four channels may be used to compare the signals.
Channel 1 is connected to the ground circuit of the ignition coil using a voltage probe.
Channel 2 is connected using a voltage probe to the supply voltage (the same wire used by the current clamp).
Channel 3 is connected using a voltage probe to the coil switching circuit from the ECU.
Channel 4 is connected using an inductive amp clamp around the positive voltage supply wire to the ignition coil. (If the pattern appears to be upside-down, turn the amp clamp around on the wire to face the other way).

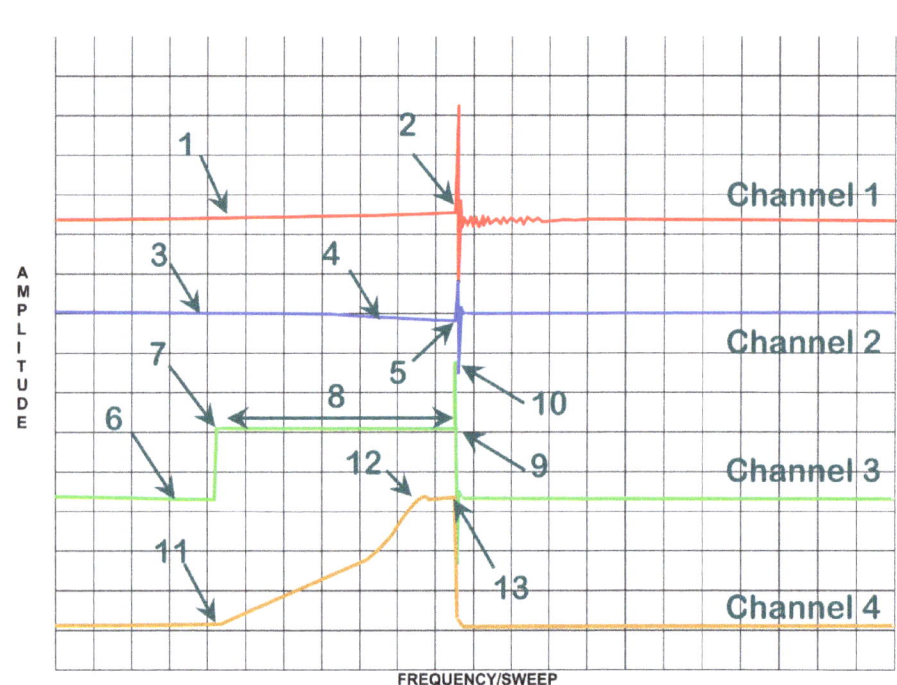
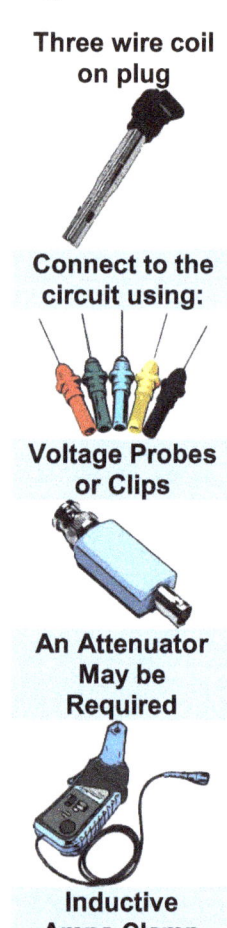

Figure 5.11 Three wire coil on plug

Automotive Ignition Systems & Waveform Analysis

Table 5.11 Waveform analysis three wire coil on plug

Waveform component	Description
1	Channel 1. With channel 1 connected to the ignition coil ground circuit, this part of the waveform should be as close to zero volts as possible (indicating a good ground).
2	Channel 1. This shows the point on the earth circuit when the ignition coil is energised. A small spike will be seen in the voltage, however, only around 0.1 of a volt would be expected; a voltage higher than this would give an indication of a bad earth. (If in doubt about the amplitude of the spike shown, compare with the other coils or a similar vehicle with no known running issues).
3	Channel 2. With channel 2 connected to the coils voltage supply, this point of the waveform should show an amplitude approximately the same as the battery operating voltage (i.e. around 14 volts with the alternator charging).
4	Channel 2. When the ignition coil is switched on, the current being drawn by the charging of the windings will cause the voltage to fall away slightly as the flow of electricity allows a **potential difference (Pd)** to occur.
5	Channel 2. This point on the waveform shows the momentary disruption to the supply voltage when the coil fires.
6	Channel 3. With channel 3 connected to the ignition coils switching circuit, the voltage should be at zero volts with the coil turned off.
7	Channel 3. As soon as the coil is switched on, the voltage will jump to around 5 volts (controlled by the ignition system ECU).
8	Channel 3. This is the 'dwell period' or charge time of the ignition coil.
9	Channel 3. This shows the point where the coil is switched off at the end of the dwell period.
10	Channel 3. The voltage spike created at this point gives an indication of the high voltage induced in the primary winding as the coil is fired.
11	Channel 4. With channel 4 connected using an amp clamp around the voltage supply wire, this will show no current flow until the coil is switched on, at which point the waveform will rise with a steady slope.
12	Channel 4. The small hump shown in the waveform at this point indicates that the coil has reached full saturation charge and the ECU is limiting the current supply to stop the windings overheating and burning out. Depending on how fast the engine is running this hump may occur earlier in the waveform (slow running) or may even disappear completely (very fast running).
13	Channel 4. As soon as the ignition coil is switched off, current flow stops immediately and drops back to zero.

Automotive Ignition Systems & Waveform Analysis

Potential difference (Pd) – a drop in electrical voltage as current is allowed to flow through a consumer once a circuit has been switched on.

Coil on plug secondary verses digital trigger

In order to check the operation of some coil on plug systems, it can be useful to compare the switching with the secondary waveform. In the following example, the primary switching is displayed on channel 1 and the secondary high-tension HT is shown on channel 2.
To display a secondary waveform, channel 2 will need the use of a high-tension inductive pick-up.

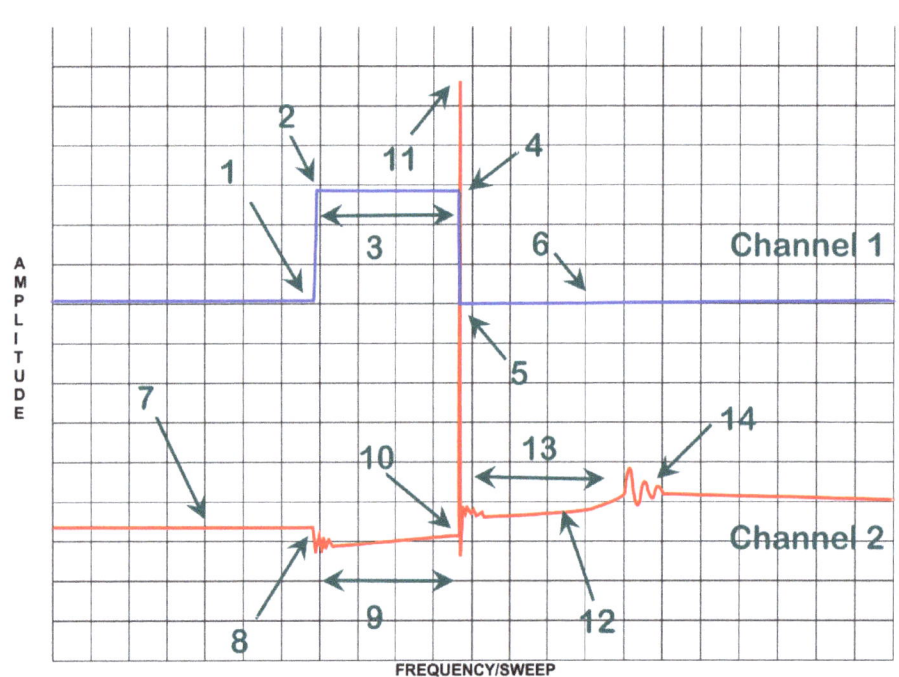

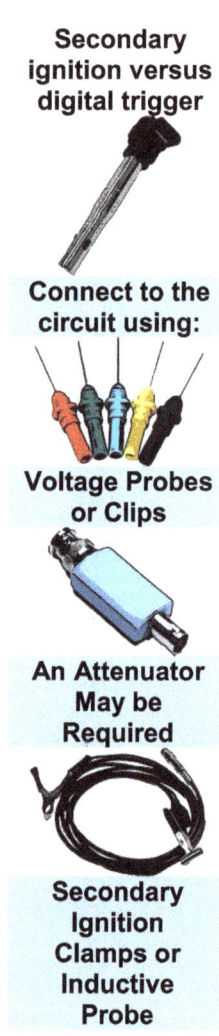

Figure 5.12 Coil on plug switching versus secondary HT

Automotive Ignition Systems & Waveform Analysis

Table 5.12 Waveform analysis secondary ignition versus digital trigger

Waveform component	Description
1	Channel 1. On channel 1, when the ignition coil is switched off, the voltage on the circuit from the ECU will show at zero.
2	Channel 1. When the ECU switches on the coil, the voltage will jump to around 5 volts (depending on manufacturer).
3	Channel 1. This part of the waveform represents the "dwell period" or switched on time of the coil.
4	Channel 1. This point shows the end of the dwell period.
5	Channel 1. As soon as the ECU breaks the circuit, the voltage immediately falls to zero.
6	Channel 1. The voltage will then remain at zero until the next coil operation.
7	Channel 2. On channel 2, the secondary circuit waveform will remain relatively flat until the primary circuit is switched on.
8	Channel 2. This is where the primary circuit switches on and the ignition coil starts to charge.
9	Channel 2. This is the 'dwell period' or charge time of the ignition coil.
10	Channel 2. This point on the waveform shows when the coil is switched off, representing the end of the dwell period.
11	Channel 2. Once the coil is switched off, the rapid collapse in the magnetic field induces a high voltage in the secondary winding. The voltage rises until it reaches a point where it can overcome the resistance of the air gap at the tip of the spark plug. The maximum amplitude achieved at this point gives an indication of how much voltage is required to fire the spark plug (shown in kilovolts KV) and should be relatively even when comparing different cylinders.
12	Channel 2. Once the initial resistance of the air gap at the tip of the plug has been overcome, it takes less voltage to maintain the spark, so the amplitude falls and a fairly shallow wave is shown sloping slightly down.
13	Channel 2. This section shows the spark line or burn time of the plug.
14	Channel 2. When the spark ends, an aftershock of coil oscillations is created by a dissipating magnetic field. The firing of the spark plug is now complete and the waveform settles out to a flat line before the next action.

Chapter 6 Automotive Network Systems and Waveform Analysis

This chapter will help you develop knowledge and understanding of automotive vehicle network systems. It will enable you to conduct effective diagnosis and repairs of system faults; supporting you by providing a breakdown of waveform images and analysing why the patterns are formed. Remember to work in a systematic way, and observe the relevant environmental, health and safety regulations at all times.

Contents

Networks	152
Communication data	158
Multiplex and networked diagnosis	162
Can Bus	163
FlexRay	165
K-Line	167
LIN Bus	168

There are many hazards associated with the service and maintenance of light vehicle electrical and electronic systems. You should always assess the risks involved with any diagnostic, maintenance or repair routine before you begin and put safety measures in place.
You need to give special consideration to the possibility of:
• The risk of electric shock.
• The hazards associated with running engines in confined spaces.
You should always use appropriate personal protective equipment (PPE) when you work on these systems. Make sure that your selection of PPE will help protect you from these hazards.

Don't forget your PPE and VPE

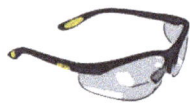

Automotive Network Systems & Waveform Analysis

Information sources

The complex nature of light vehicle electric and electronic systems requires a good source of technical information and data. In order to conduct diagnostic, maintenance and repair procedures, you need to gather as much information as possible before you start.
Sources of information may include:

Table 6.1 Possible information sources

Verbal information from the driver	Vehicle identification numbers
Service and repair history	Warranty information
Vehicle handbook	Technical data manuals
Workshop manuals/Wiring diagrams	Safety recall sheets
Manufacturer specific information	Information bulletins
Technical helplines	Advice from other technicians/colleagues
Internet	Parts suppliers/catalogues
Jobcards	Diagnostic trouble codes
Oscilloscope waveforms	On vehicle warning labels/stickers
On vehicle displays	Reference/Textbooks

Always compare the results of any inspection, testing or diagnosis to suitable sources of data. Remember that no matter which information or data source you use, it is important to evaluate how useful and reliable it will be to your safety, diagnostic, maintenance and repair routine.

Where to start?

Step 1 • A good systematic diagnostic routine should always begin with careful questioning of the driver to gather as much information about the symptoms and history of the fault as possible.

Step 2 • After a brief visual inspection for obvious signs of damage or safety issues, the system should be tested to try and recreate the fault.

Step 3 • The vehicle should then be scanned for diagnostic trouble codes and any codes should be recorded. (If possible, a full scan should be conducted, as issues in unrelated systems can sometimes affect the operation of others).

Step 4 • Any codes should be cleared and the vehicle should be tested over a complete drive cycle.

Step 5 • Rescan the vehicle and concentrate diagnosis around any codes that have returned.

Step 6 • Connect the oscilloscope to the suspected circuit and analyse any waveforms produced.

Automotive Network Systems & Waveform Analysis

Networks

As computerised control of vehicles increased, it was discovered that it was more efficient to have separate electronic control units and join them in a network which shared information. The communication between ECU's is similar to parallel processing used in computers, meaning the workload can be shared and prioritised.
Not only has networking brought speed and cost benefits to the production of vehicles, but the reliability of system operation has also improved. Most manufacturers now have their own network design and some of the more common types are described in the next section.

CAN Bus

This is probably one of the most widely used networks within vehicle design, and the name 'CAN Bus' has become synonymous with ECU communication to the point where it is often used to describe all in car networking, even if another type is actually being used.
CAN Bus is a network communication standard where information is bundled into a 'data packet' and sent onto the Bus system along two twisted wires. Every **node** on the Bus system receives the message and acts if required.

Node – the name used to describe individual ECU's in a vehicle network system.

Arbitration – a set of rules used to prioritise network communication for importance.

FlexRay

FlexRay is a network communication system designed to be faster and more reliable than CAN Bus. The data frames are clock synchronised with speeds of up to 10Mbps and the system can operate with a reduced bandwidth, even if one channel fails.

LIN Bus

LIN or Local Interconnect Network, was designed and introduced to be a cheaper alternative to other network systems. Using a single wire, all messages are sent from one master node, and all other nodes in the system are slaves. Because all messages are sent from the master to the slave, there is no data collision and therefore no **arbitration** is required. The system operates in two modes, 'awake' and 'sleep'. If no activity is seen on the network after a predetermined time period, the nodes will enter sleep mode. When required the master node will send a 'wake up' data frame.

MOST

MOST or Media Oriented Systems Transport is a very high speed network used to transmit and receive large amounts of data, particularly audio, voice and video, between different units within the car. Unlike some of the other networks available, MOST uses plastic optical fibre (POF) and light emitting diodes LED to transmit the signals. MOST is currently the global standard used in vehicles for transmitting multimedia and is not directly testable using an oscilloscope.

Automotive Network Systems & Waveform Analysis

Network communication

As the amount of technology on cars has increased, demand for faster computer operation and processing has also risen. Advances in vehicle management include:

- Engine management.
- Body control.
- Chassis systems.
- Transmission.
- Infotainment.
- Traction control.
- Safety systems.

ECU's were becoming bigger to cope with system requirements, and large amounts of wiring were needed to distribute electrical power around the car. These demands also generated a rise in the number of sensors required, leading to complication, extra weight and increases in the cost of manufacture.

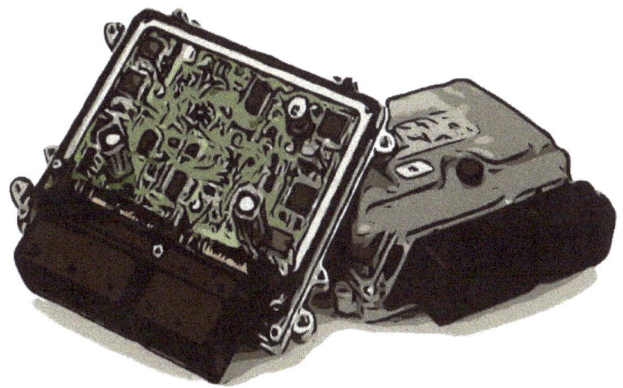

Figure 6.1 Electronic control unit ECU

The reason why ECU's were becoming larger and larger was due to the need for more connections and pins where they joined the wiring. There was a limit to how small these connections could be made, and how closely the wires in the loom could be bundled. Multiplexing has allowed a reduction in sensors and wiring, as it allows the ECU's to share information on a network.

To reduce the amount of sensors and wiring needed for system operation, **multiplexing** was introduced. Multiplexing simply means carrying out more than one operation at a time. Multiplex networking brings with it the benefits of efficient control of vehicle systems, it gives manufacturers the opportunity to configure vehicle options late in the manufacturing process and the ability to update software to overcome running issues and fix bugs once the vehicle has been launched.

Instead of a single large ECU in a vehicle, smaller ECU's were developed that managed individual systems. These single ECU's became known as **nodes**. The nodes are connected to each other by a communication wire, which allows information to be shared in a **network**. When one of the ECU's receives information from a sensor, it processes the signal and acts if required. It then passes on this information to the communication network wire linking the other ECU's, which then use that information if required, and once again pass it on. This means that signals from a single sensor can be shared across a number of different vehicle systems.

Multiplexing – a method of carrying out more than one operation at once.

Nodes – ECU's connected to a computer network (from the Latin word 'nodus', which means knot).

Automotive Network Systems & Waveform Analysis

Network – a number of computers connected together so that they are able to communicate with each other.

Topology diagrams – the name given to the specialist wiring diagrams created to identify the major components of an in-vehicle network system.

Physical layer

The name given to the wiring of an in-vehicle network system is the physical layer. It is used to join the various nodes/components to each other, and also to connect networks of different speeds and systems. In order to help locate and trace the network systems, wiring schematics known as **Topology diagrams** are created by manufacturers, showing the layout of the major components.

It is often possible to conduct an initial oscilloscope diagnosis of network systems at the pins of the vehicle data link connector. Due to the standardised layout of the 16-pin connector the terminals can be identified from the image shown below:

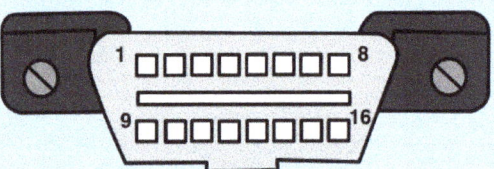

1. Manufacturer specific (sometimes used for network communication).
2. Bus positive SAE J1850 PWM and VPW.
3. Manufacturer specific (sometimes used for network communication).
4. Chassis ground.
5. Signal ground.
6. CAN High.
7. K-Line of ISO9141-2 and ISO14230-4.
8. Manufacturer specific (sometimes used for network communication).
9. Manufacturer specific (sometimes used for network communication).
10. Bus negative SAE J1850 PWM.
11. Manufacturer specific (sometimes used for network communication).
12. Manufacturer specific (sometimes used for network communication).
13. Manufacturer specific (sometimes used for network communication).
14. CAN Low.
15. L-Line of ISO9141-2 and ISO14230-4.
16. Battery voltage.

Automotive Network Systems
& Waveform Analysis

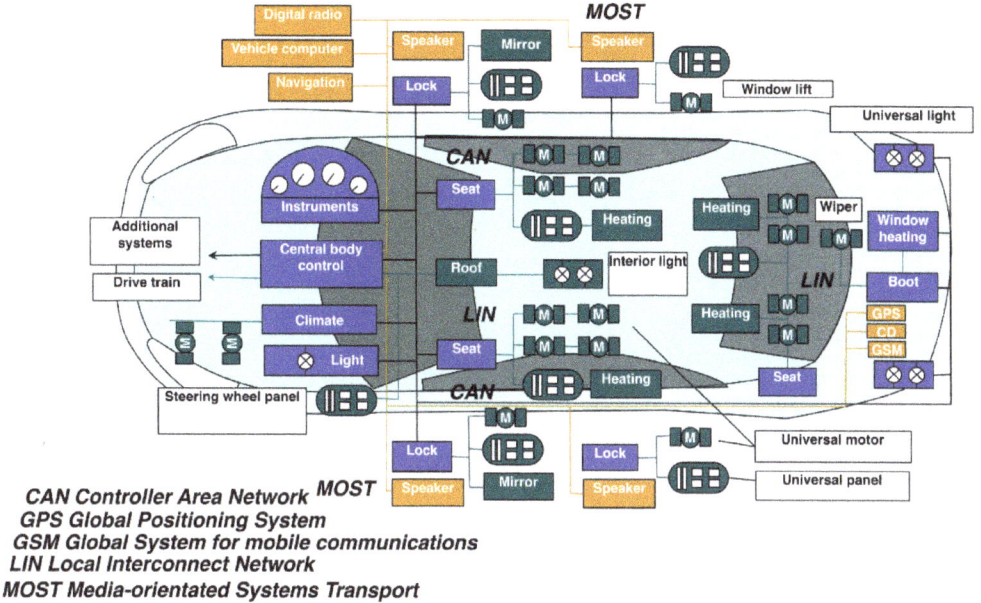

Figure 6.2 In-vehicle network systems

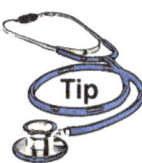 A common cause of network faults and excessive interference is badly fitted aftermarket electrical equipment. It is good practice to examine the vehicle for non-standard equipment, and if necessary, systematically disconnect components while examining live waveforms to help locate the cause of any issues.

CAN Bus

There are a number of different network types and manufacturers available, but the name CAN Bus has been adopted by many technicians to describe nearly all network systems.
Controller area network (CAN) was introduced by Robert Bosch in the 1980s and is an international standards organisation (ISO) standard for a serial multiplex communication protocol.
The advantages of CAN Bus are:

- Transmission speeds are much faster than those used in conventional communication (up to 1 Mbps), allowing much more data to be sent.
- The system is very immune to interference (noise), and the data obtained from each error detection device is more reliable.
- Each ECU connected via the CAN, communicates independently, therefore if an ECU is damaged or faulty, communications can be continued in many cases.

Automotive Network Systems
& Waveform Analysis

The CAN Bus line consists of two cables, known as CAN L and CAN H (CAN Low and CAN High, respectively). The CAN High and Low wires are twisted together and this helps to cancel out noise which may be caused by electromagnetic interference from other vehicle electrical systems. At the ends of each CAN line are terminal resistors that help to dampen out voltage spikes (**back EMF**) which could be caused as the communication is triggered on and off. The CAN Bus lines connecting two dominant ECU's are the main Bus lines, and the CAN Bus lines connecting each individual ECU are the sub-Bus lines. Each ECU communicates with the CAN periodically sending information from several sensors. This information is circulated on the CAN Bus as a data packet. Each ECU needing data on the CAN Bus can receive these data frames sent from each ECU simultaneously. A single ECU transmits multiple data frames. When data packets conflict with one another (when more than one ECU transmits signals at the same time), data is prioritised for transmission by a process called **'mediation'**.
If mediation is required:

1. The data frame with high priority is transmitted first according to ID codes embedded in the data packet.
2. Transmission of low-priority data is suspended by the issuing ECU until the Bus clears (when no transmission data exists on the CAN Bus).
3. The ECU containing suspended data frames transmits when the Bus becomes available.

If the suspended state continues for a specified time, new data (data packet content) is created and sent and this can be seen as a very "crowded" waveform on an oscilloscope.

The nodes are connected by a single communication line, which allows the exchange of multiple pieces of data. The communication line can link these ECU's in the following layouts:

- A large loop or ring, known as a daisy chain.
- A star pattern, known as a server system.
- Connected in parallel to a single Bus line.

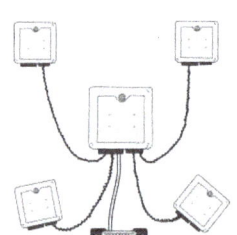
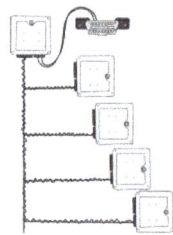

Figure 6.3 Vehicle network layouts

When a daisy chain layout is used, the data sent travels in both directions at once, which gives much greater reliability. If one wire is damaged or broken within the loop, the information can still arrive at the appropriate ECU as it comes from the other direction. The data is not only more reliable but this system also improves malfunction diagnosis.

Automotive Network Systems & Waveform Analysis

Back EMF – this is a voltage spike (a reverse electromotive force) created when a circuit is switched off.

Mediation – a negotiation to resolve difficulties or conflict. In multiplex it is the rule that sets the priority of messages.

Gateway – an ECU that joins two different system/speed Busses together and translates/transmits the data packets into a format that can be used on the different part of the network.

Communication wires

The wiring for network communication can be constructed from three main methods:

- Coaxial wiring.
- Twisted copper wiring.
- Fibre optical.

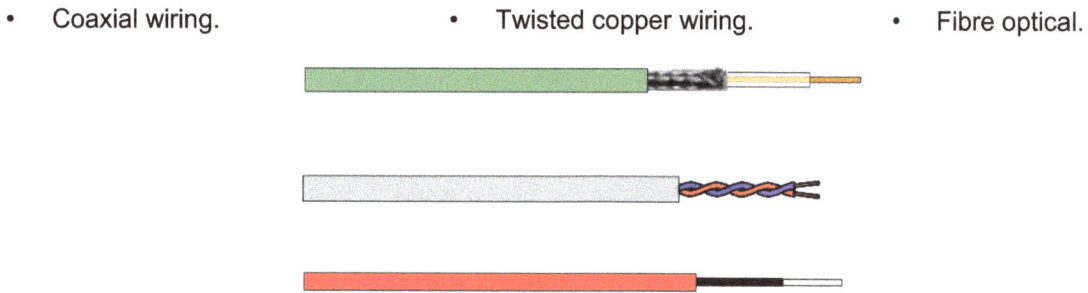

Figure 6.4 Network communication wire types

It is possible to use a mixture of layout types and communication wires for different vehicle systems (i.e. powertrain, chassis, body and infotainment). These systems can then be connected to each other through individual electronic control units known as **'gateways'**. Gateways are used in order for Busses of different speeds to co-operate. The gateway is employed to link the Busses, and it is used to encode and decode messages. The gateway may be a separate unit or it may be incorporated in an ECU. The gateways allow the different networks to share information by translating between different Bus communication speeds. When the networks are shown laid out on a diagram, it is known as the 'physical layer' or 'topology'.

The word 'Bus' is used in various situations. One meaning of Bus is a vehicle that collects you from one place and delivers you to another. This is very similar to its meaning within a communication network. Information is picked up at one point on the communication line, it then takes a route around the system and stops at various ECU's (like Bus stops).

Automotive Network Systems & Waveform Analysis

Communication data

When an ECU receives a signal from a vehicle sensor, it processes this and places the information on the network Bus as a data packet. The data packet is usually made up of the following:

- A header, **SOF (Start of Frame)**: the equivalent of 'hello, I am transmitting a message'.
- The priority **ID (Identifier) region**: how important this message is, e.g. vehicle safety information will be more important than a bodywork communication such as a command to open an electric window.
- Data length **Control region**: this is so the receiver knows it has not lost or 'misheard' any of the information.
- Data type **Data region**: what type of information is contained, e.g. speed, temperature, etc.
- Data **Data region**: the actual sensor information itself.
- An error detection code **Cyclic Redundancy Check (CRC) region**: this says 'has all the information been received?'
- End of message **EOF (End of Frame)**: 'goodbye'.
- Finally, a request for a response from the receiving ECU **ACK (Acknowledge) region**: this says 'thank you, I got your message'.

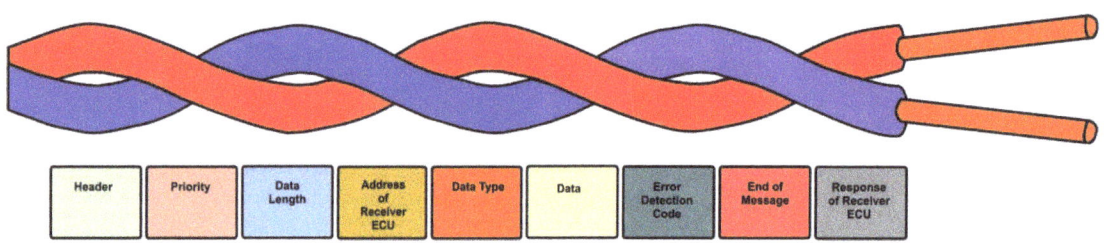

Figure 6.5 CAN Bus communication data packet

To help reduce the possibility of data **corruption** caused by misinterpretation or external electromagnetic interference, a CAN Bus system uses two communication wires instead of one, twisted over and over each other in a spiral. The same data is sent on both of these communication wires as an on and off voltage signal. One signal is sent as a positive switch and one is sent as a negative switch, providing a mirror image on each network wire, which are known as CAN High and CAN Low. The **potential difference** between the voltages on the two lines produces a digital signal that can be processed into information.

If the oscilloscope patterns from CAN High and Low don't line up, check the channel speeds of your scope before assuming that there is a fault with the system.

Automotive Network Systems & Waveform Analysis

DATA TRANSMISSION - High Speed

The transmitting ECU sends switched voltage through the CAN H and CAN L Bus.
2.5 to 3.5 V signals to the CAN High line.
2.5 to 1.5 V signals to the CAN Low line.
The receiving ECU reads the data from the CAN lines as potential difference of between 3.5 and 1.5 volts.

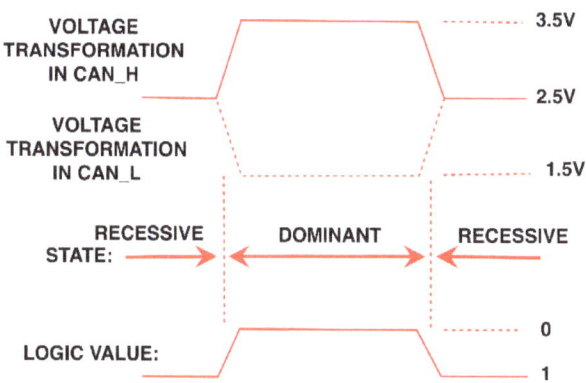

Figure 6.6 Data transmission high speed

In Figure 6.6, 'Recessive' refers to the state where both CAN H and CAN L are under the 2.5 V state, and 'Dominant, refers to the state where CAN H is under the 3.5 V state and CAN L is under the 1.5 V state. These values correspond to a binary value of either 1 or 0.
Recessive = Logic value of 1
Dominant = Logic value of 0

DATA TRANSMISSION - Low Speed

The transmitting ECU sends switched voltage through the CAN H and CAN L Bus.
0 to 4 V signals to the CAN High line.
1 to 5 V signals to the CAN Low line.
The receiving ECU reads the data from the CAN High and CAN Low as potential difference of between 5 and 0 volts.

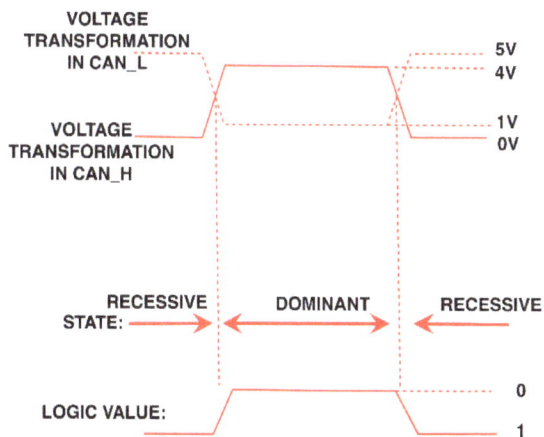

Figure 6.7 Data transmission low speed

In Figure 6.7 'Recessive' refers to the state where CAN H is at 0V and CAN L is at 5V
'Dominant' refers to the state where CAN H is at 4V state and CAN L is 1V.
Recessive = Logic value of 1
Dominant = Logic value of 0

The communication wires are exactly the same length, and as the data travels at the same speed, both versions of the information should arrive at the receiver at the same time. These messages can now be compared to help identify data corruption. The opposing voltage of the signals transmitted down the communication wires will also help cancel out electromagnetic interference from other systems.

Automotive Network Systems & Waveform Analysis

Terminating resistances

Another factor that might affect a CAN Bus system, creating data corruption, are network voltage spikes caused by back EMF during switching. A 120 ohm resistor is connected across the CAN High and CAN Low circuits at each end which help to reduce voltage spikes and lower the possibility of corruption; they act like shock absorbers for electricity.

Termination resistances can give a good indication of correct circuit operation.
If an ohmmeter is connected in parallel across CAN High and CAN Low (using pins 6 and 14 of the data link connector for example) the total recorded resistance will be halved.

- If 60Ω is shown CAN High and CAN Low should be OK.
- If O/L (Infinity is shown) an open circuit exists in both lines.
- If 0Ω is shown a dead short exists.
- If 120Ω is shown one CAN line may be at fault (confirm communication using Oscilloscope).

Corruption – a breakdown of integrity or communication.

Potential difference – the difference between two voltage values.

Infotainment – a combined information and entertainment system.

Hexadecimal – a method of counting in batches of sixteen that can be used as a computer programming language.

Bus speeds

There are three main CAN Bus speeds:

- Low speed: used for instrumentation, body control and comfort, etc.
- It operates at a rate of 33,000 bits of information per second (33kbps).

- High speed: used for powertrain control and safety critical information, etc.
- It operates at a rate of 500,000 bits of information per second (500 kbps).

- Very high speed: used for high volumes of data transmission in **infotainment** systems (such as streaming video and music), etc.
- It will operate at a rate of 25,000,000 bits of information per second (25 Mbps).

This is sometimes known as baud rate, but only if related to binary (1 & 0). Modern computer communication may have more than 2 symbol states (**hexadecimal** programming language) so baud rate can be misleading.

Automotive Network Systems & Waveform Analysis

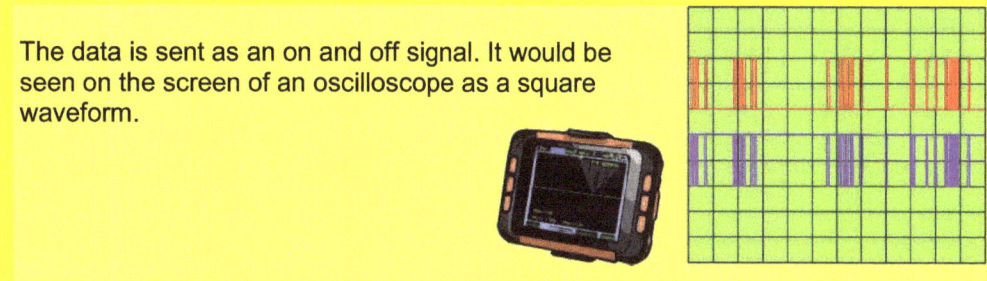

The data is sent as an on and off signal. It would be seen on the screen of an oscilloscope as a square waveform.

The main CAN Bus lines (High and Low) can often be tested with an oscilloscope by connecting to pins 6 and 14 of the 16 pin diagnostic socket.

System reliability

The CAN Bus system is more reliable than a standard wiring system due to the fact that a single open Bus wire would not stop communication. Two open Bus wires can stop communication, but as more ECU's are used for control only part of the system may fail.

Short circuits can have a catastrophic effect on network communication. A short to either positive or earth will disrupt the communication on the Bus wire, as an on and off signal can no longer be transmitted. If viewed on and oscilloscope screen, this would be a flat line at either 0V or 5V (this can also be around 12V if shorted to battery). To avoid total failure of the system, Bus-cut relays can be used. These are a type of circuit breaker that isolate part of the network, allowing the rest of the system to continue communicating.

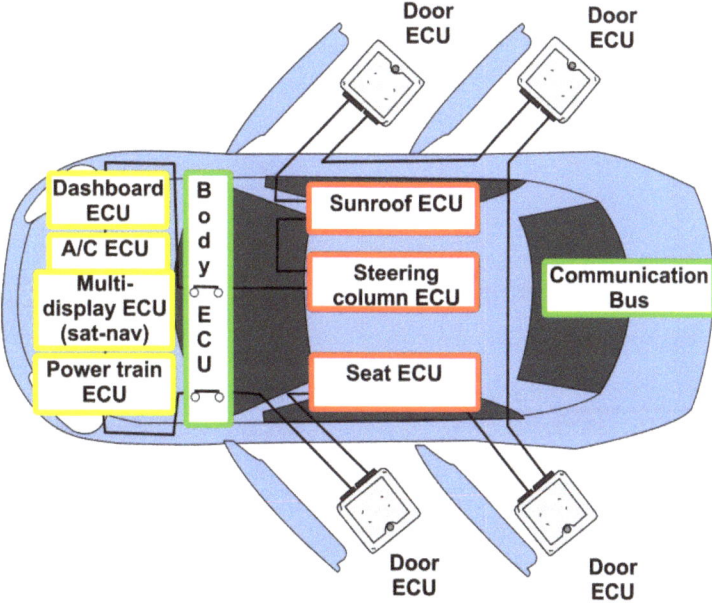

Figure 6.8 A daisy chain network layout with Bus cut relays

Automotive Network Systems & Waveform Analysis

Tip: It is important to remember that when diagnosing in-vehicle network systems, that operating voltages should be kept at or above battery voltage during your tests. Different systems and ECU's will have varying operating thresholds, and if voltages drop below specified values, some may switch off or go to sleep.
Intermittent communication issues could be caused by battery or charging system faults, so these should be thoroughly checked as part of your systematic diagnostic routine.

Multiplex and networked diagnosis

If a critical network failure occurs, such as short to positive or earth, the vehicle may suffer a complete communication loss.

With a networked system, if communication is lost within a certain area, number of items will not work and numerous trouble codes may be generated. Having connected a scan tool and retrieved the diagnostic trouble codes, you should look for the code that is the root cause. Communication failures are normally an effect of the original fault (i.e. 'unable to communicate' or 'communication lost'). You should ask yourself, 'Is this the cause or an effect created by the fault?' CAN Bus systems report communication faults as live data. As a result, once you have identified the cause trouble code, you should be able to conduct a diagnosis by disconnecting and isolating components or sections of wiring loom until communication is re-established.

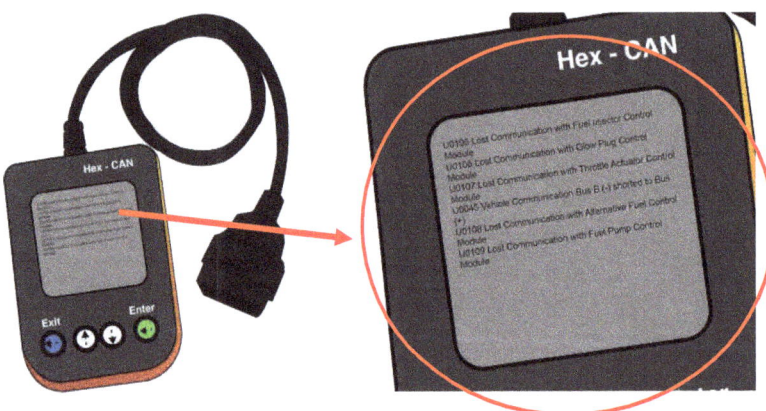

Figure 6.9 CAN Bus diagnostic trouble codes

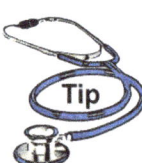

Tip: CAN Bus faults may cause issues on non-related systems, always do a full system scan where possible.

Automotive Network Systems & Waveform Analysis

Once diagnostic trouble codes have been obtained indicating a fault with a networked system, it will be necessary to connect an oscilloscope to help further investigate the possible fault. The following section shows different types of network communication and how to interpret the waveforms produced.

Can Bus

One of the most popular designs used in vehicle networking is CAN Bus (Controller Area Network). This system can often be identified as a pair of twisted wires entering or leaving an ECU. An oscilloscope can be connected to these wires by 'back probing' at the ECU socket, but if the vehicle is operating using CAN Bus, a good place to connect to the main circuit is at pins 6 and 14 of the diagnostic socket (DLC).

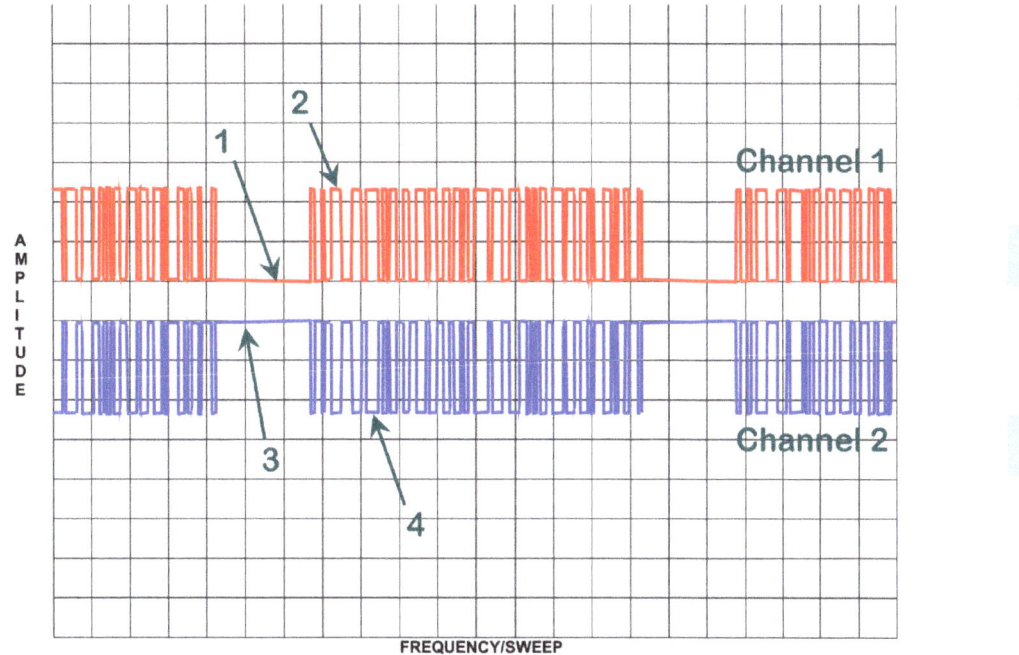

Figure 6.10 CAN Bus (CAN H and CAN L)

Table 6.10 Waveform analysis CAN High and CAN Low

Waveform component	Description
1	Channel 1. Channel 1 is connected to CAN H (High) as it switches positively. This shows a voltage of 0 or 2.5 volts in the off position depending on network speed.
2	Channel 1. When switched on the voltage will jump to 3.5 or 4 volts depending on network speed.
3	Channel 2. Channel 2 is connected to CAN L (Low) as it switches negatively. This shows a voltage of 5 or 2.5 volts in the off position depending on network speed.
4	Channel 2. When switched on the voltage will fall to 1 or 1.5 volts depending on network speed.

Automotive Network Systems & Waveform Analysis

By changing the frequency/sweep on the oscilloscope and aligning the voltage amplitudes between channel 1 and channel 2 it is possible to compare the two patterns and see the potential difference from CAN High and CAN Low. This can then be interpreted as a dominant or recessive logic value. (See Figures 6.6 and 6.7).
Recessive = Logic value of 1
Dominant = Logic value of 0

Tip: It is important when checking the waveform, that the patterns from CAN High and Low show equal and opposite with clean edges. This gives an indication that the network wiring circuit is operating effectively and if an individual ECU is not responding, it is probably caused by the ECU itself.

Safety: Never use insulation piercing probes to measure CAN Bus signals as this can affect the integrity of the wiring and cause communication issues.

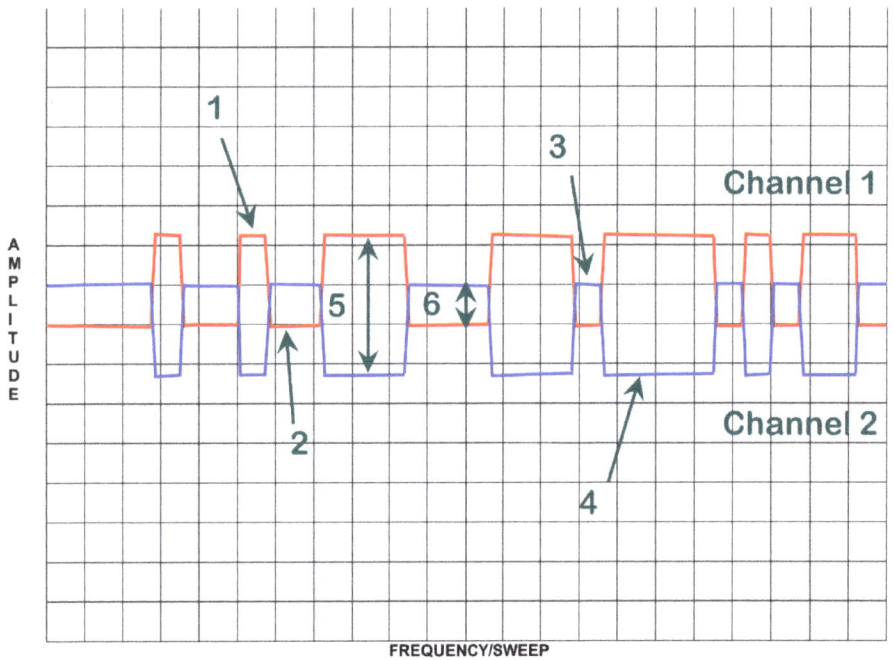

Figure 6.11 CAN Bus potential difference

Automotive Network Systems & Waveform Analysis

Table 6.11 Waveform analysis CAN Bus potential difference

Waveform component	Description
1	Channel 1. Channel 1 is connected to CAN H (High) as it switches positively. This shows a voltage of 3.5 or 4 volts in the on position depending on network speed.
2	Channel 1. When switched off the voltage will fall to 0 or 2.5 volts depending on network speed.
3	Channel 2. Channel 2 is connected to CAN L (Low) as it switches negatively. This shows a voltage of 5 or 2.5 volts in the off position depending on network speed.
4	Channel 2. When switched on the voltage will fall to 1 or 1.5 volts depending on network speed.
5	This section of the waveform shows a dominant logic value of 0.
6	This section of the waveform shows a recessive logic value of 1.

FlexRay

As the demand for information transmission in vehicle systems has grown, alternative network designs have been produced that can handle large amounts of data. An example of this is FlexRay.

Developed by a consortium of manufacturers (which disbanded in 2009), it can handle Bus speeds of up to 10Mbps. When checking signals using an oscilloscope, it can be viewed in a similar manner to Can Bus.

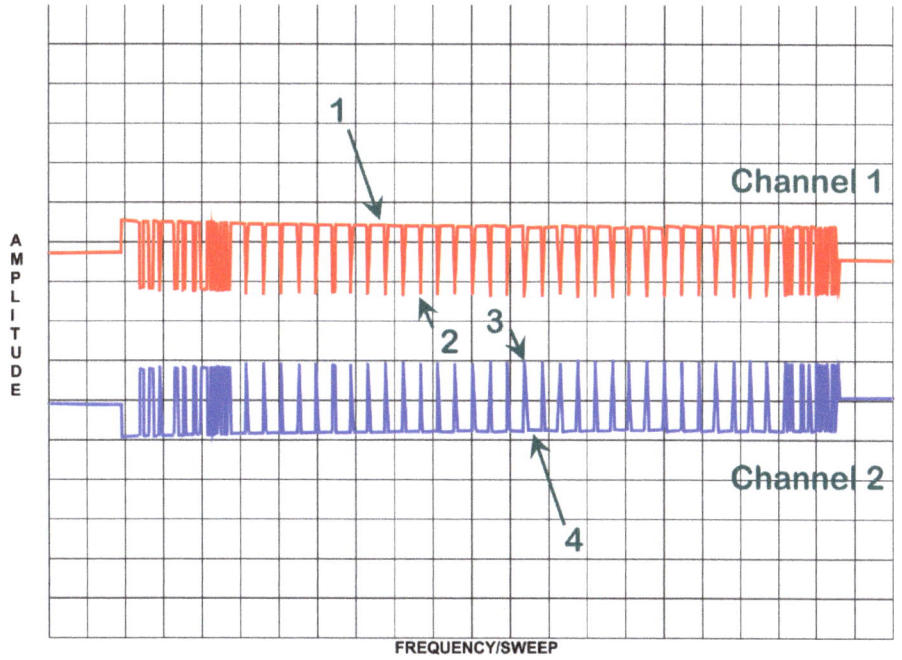

Figure 6.12 FlexRay waveform data

Automotive Network Systems & Waveform Analysis

Table 6.12 Waveform analysis FlexRay data

Waveform component	Description
1	Channel 1. Shows positively switched data with this section as on.
2	Channel 1. Shows positively switched data with this section as off.
3	Channel 2. Shows negatively switched data with this section as off.
4	Channel 2. Shows negatively switched data with this section as on.

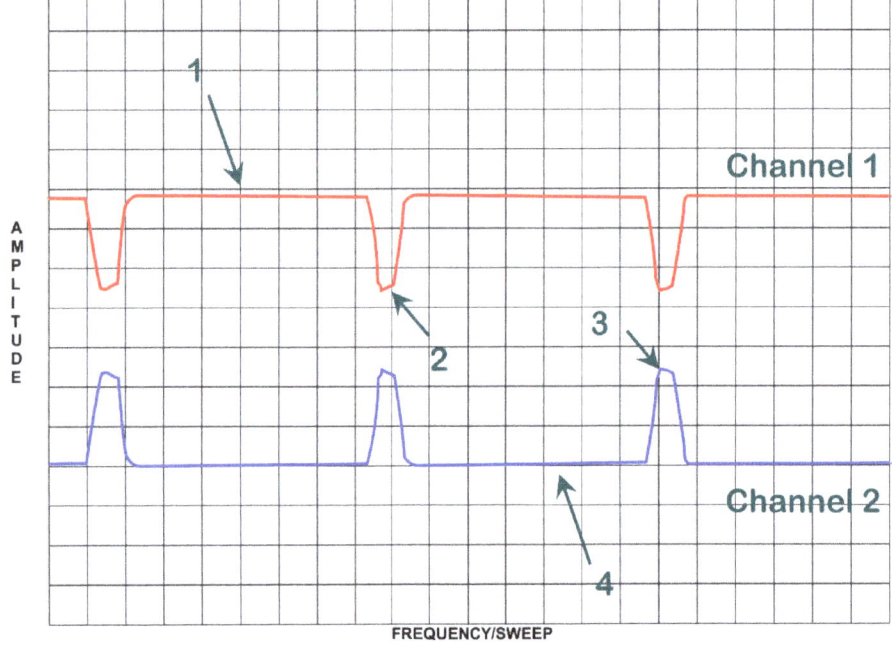

Figure 6.13 FlexRay waveform (zoomed)

Table 6.13 Waveform analysis FlexRay (zoomed)

Waveform component	Description
1	Channel 1. Shows positively switched data with this section as on.
2	Channel 1. Shows positively switched data with this section as off.
3	Channel 2. Shows negatively switched data with this section as off.
4	Channel 2. Shows negatively switched data with this section as on.

Automotive Network Systems & Waveform Analysis

K-Line

K-line is a low speed, single wire network communication which, unlike CAN Bus, does not have a central or primary ECU controlling the network. This allows all ECU's to send and receive information equally. K-Line is designed to support diagnostic equipment, sending information between ECU's and the diagnostic socket, which can then be read by a scan tool. As there is only one wire used with this design, the signals are transmitted using a pulsed on and off voltage which is then translated into a binary code.

K-Line data can be checked at the diagnostic socket by connecting the oscilloscope to pin 7 of the DLC.

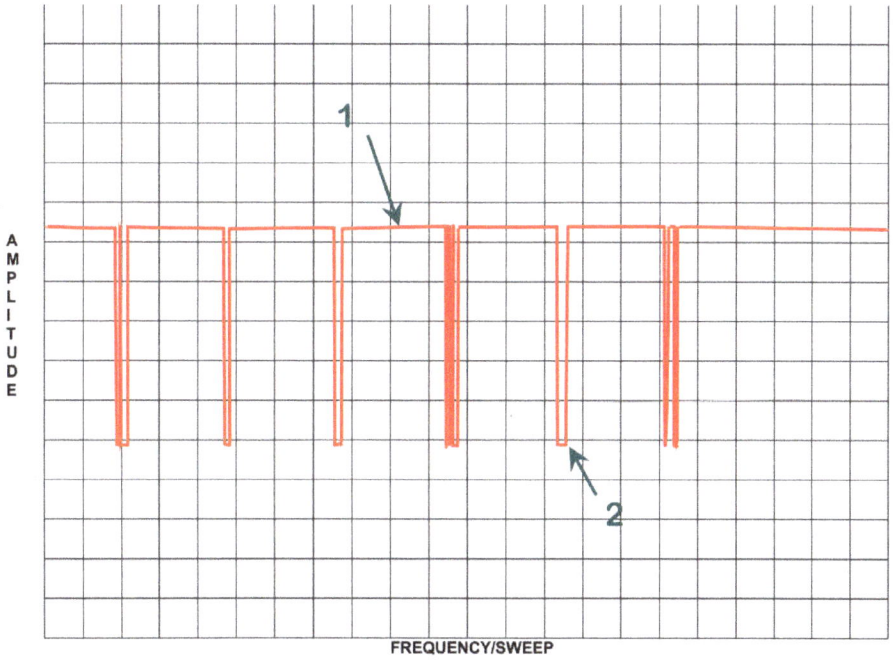

Figure 6.14 K-Line waveform data

Table 6.14 Waveform analysis K-Line data

Waveform component	Description
1	This shows a pulsed reading of 12 volts with a digital logic value of 0.
2	This shows a pulsed reading of 0 volts with a digital logic value of 1.

Automotive Network Systems & Waveform Analysis

Unlike CAN Bus and FlexRay, communication signals are sometimes broken down into sections and sent individually. (Other systems send the entire communication in one data packet).

LIN Bus

LIN Bus or Local Interconnect Network is a low cost single wire alternative to some of the more expensive network types. It was specifically developed to allow the use of cheap components, while maintaining serial data communication between electronic control units. Its basic design consists of a master ECU and typically up to 15 slave ECU's, and because the master sends the messages, there should in theory be no collision of data. Messages sent on a LIN Bus system can be seen on an oscilloscope as a square waveform, representing the binary states of 1 and 0.

The master ECU of a LIN Bus network system can also act like a slave unit by replying to its own messages.

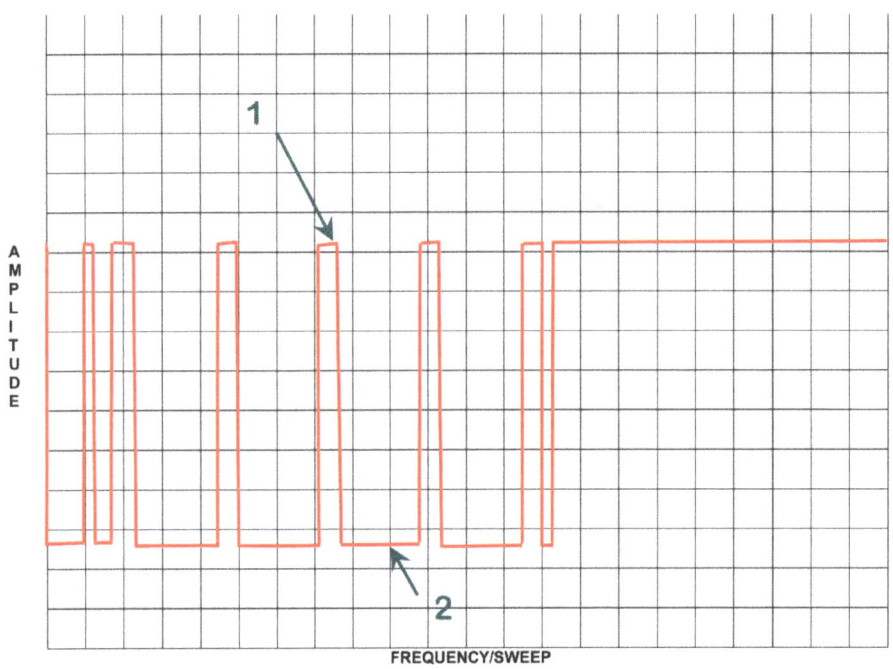

Figure 6.15 LIN Bus waveform data

Automotive Network Systems & Waveform Analysis

Table 6.15 Waveform analysis LIN Bus data

Waveform component	Description
1	This section of the waveform represents an off signal and shows a recessive logic state of 1. It will typically have a voltage of around 80% of battery operating voltage and can be slightly higher or lower, depending if the engine is running and charging, or not. The waveform should be free of obvious spikes or distortion.
2	This section of the waveform represents an on signal and shows a dominant logic state of 0. It will typically have a voltage of around 20% of zero volts and will often be approximately 1 volt. The waveform should be free of obvious spikes or distortion.

LIN Bus serial data cannot be decoded using an oscilloscope, so the waveform should be checked for quality and that it is not affected by moving the wiring harness or ECU plugs.

Automotive Pressure Testing & Waveform Analysis

Chapter 7 Automotive Pressure Testing and Waveform Analysis

This chapter will help you develop knowledge and understanding of automotive mechanical pressure testing using a transducer and oscilloscope. It will enable you to conduct effective diagnosis and repairs of system faults; supporting you by providing a breakdown of waveform images and analysing why the patterns are formed. Remember to work in a systematic way, and observe the relevant environmental, health and safety regulations at all times.

Contents
Pressure transducer testing 172

Petrol intake manifold pressure 173

Exhaust pressure 178

In-cylinder compression 181

Crankcase pressure 184

Fuel pressure 186

There are many hazards associated with the service and maintenance of light vehicle mechanical systems. You should always assess the risks involved with any diagnostic, maintenance or repair routine before you begin and put safety measures in place.
You need to give special consideration to the possibility of:
• The hazards associated with running engines in confined spaces.
You should always use appropriate personal protective equipment (PPE) when you work on these systems. Make sure that your selection of PPE will help protect you from these hazards.

Don't forget your PPE and VPE

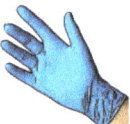

Automotive Pressure Testing & Waveform Analysis

Information sources

The complex nature of light vehicle systems requires a good source of technical information and data. In order to conduct diagnostic, maintenance and repair procedures, you need to gather as much information as possible before you start.
Sources of information may include:

Table 7.1 Possible information sources

Verbal information from the driver	Vehicle identification numbers
Service and repair history	Warranty information
Vehicle handbook	Technical data manuals
Workshop manuals/Wiring diagrams	Safety recall sheets
Manufacturer specific information	Information bulletins
Technical helplines	Advice from other technicians/colleagues
Internet	Parts suppliers/catalogues
Jobcards	Diagnostic trouble codes
Oscilloscope waveforms	On vehicle warning labels/stickers
On vehicle displays	Reference/Textbooks

Always compare the results of any inspection, testing or diagnosis to suitable sources of data. Remember that no matter which information or data source you use, it is important to evaluate how useful and reliable it will be to your safety, diagnostic, maintenance and repair routine.

Where to start?

Step 1
- A good systematic diagnostic routine should always begin with careful questioning of the driver to gather as much information about the symptoms and history of the fault as possible.

Step 2
- After a brief visual inspection for obvious signs of damage or safety issues, the system should be tested to try and recreate the fault.

Step 3
- The vehicle should then be scanned for diagnostic trouble codes and any codes should be recorded. (If possible, a full scan should be conducted, as issues in unrelated systems can sometimes affect the operation of others).

Step 4
- Any codes should be cleared and the vehicle should be tested over a complete drive cycle.

Step 5
- Rescan the vehicle and concentrate diagnosis around any codes that have returned.

Step 6
- Connect the oscilloscope using a suitable pressure transducer and analyse any waveforms produced.

Automotive Pressure Testing & Waveform Analysis

Pressure transducer testing

Apart from electrical and electronic diagnosis, an oscilloscope can also be used for mechanical testing. Some equipment manufacturers produce an adapter that acts as a pressure **transducer**, which can be attached to the scope, and convert pressure or vacuum readings into a diagnosable waveform. These adapters are often powered by a rechargeable battery, and when attached to the oscilloscope, show pressure as amplitude on the Y axis and frequency or speed across the X axis.

Transducer – a device that converts one form of energy to another; pressure into an electrical signal for example.

A pressure transducer adapter should only be used/connected to recommended systems that operate within the manufacturing tolerance of the equipment. Exceeding the measurable/recommended pressures can lead to personal injury, vehicle and equipment damage.

Pressure sensors

Pressure sensors are used in various systems to help provide vehicle system monitoring. The next section gives a brief description of some of the most common applications.

Intake

Inlet manifold pressure sensors or manifold absolute pressure sensors (MAP) are connected to the intake and help to measure manifold depression. Measurement of manifold depression will give an indication of engine load and air inducted into the engine. This allows the ECU to calculate the correct quantity of fuel to inject, depending on operating conditions and maintain air/fuel ratios.

Exhaust

Since the introduction of Diesel particulate filters (DPF) and complex emission reduction systems, it has become important for the engine management system to monitor exhaust pressure. These systems have made it necessary to add another step to many diagnostic routines. Exhaust pressure measurement will give an indication of the engines ability to breathe and this information is vital in assessing mechanical health.

In cylinder

A pressure transducer mounted in the cylinder of an engine is another effective method of assessing engine mechanical health. Pressures that are too high may lead to an uncontrolled detonation of fuel and therefore a misfire, pressures that are too low will reduce mechanical efficiency.

Automotive Pressure Testing & Waveform Analysis

Crankcase

If crankcase pressure is monitored, this information can be used to diagnose worn pistons and cylinders. Excessive crankcase pressure is mainly caused through a process known as 'blow-by' as compression and combustion pressures leak past the piston rings and into the crankcase.

Petrol intake manifold pressure cranking

With the pressure transducer connected to the oscilloscope, adapters can be used to attach it to the intake manifold of a petrol engine via a suitable vacuum pipe connection; this should be after the throttle butterfly. With the engine being cranked, the intake pulses can then be analysed.

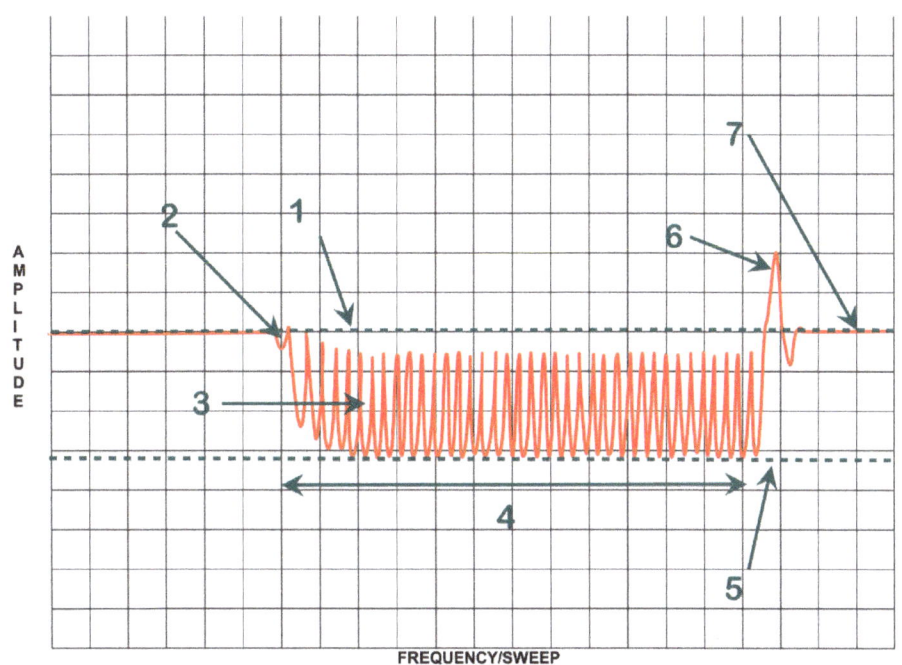

Figure 7.2 Intake manifold pressure while cranking

It is important to remember that the pressure pulsations in the manifold will be affected by the position of the throttle butterfly.
While this test is being conducted, it is recommended that the throttle be held in one position, so a reliable assessment of the waveform can be obtained.

Automotive Pressure Testing & Waveform Analysis

Table 7.2 Waveform analysis Intake manifold pressure (cranking)

Waveform component	Description
1	The cursor line shown at this point represents atmospheric pressure.
2	The first pulse indicated on the waveform shows the start of cranking.
3	The waveform pulsations here, show the intake depressions created by the strokes of the pistons; they should all be relatively even. A pulsation that doesn't pull as low as the others may indicate a poor compression. A peak that is higher than the others may indicate a leaking intake valve, allowing cylinder pressure back into the manifold.
4	This shows the length of time that the engine is being cranked.
5	The cursor line here shows the maximum intake manifold vacuum achieved during cranking.
6	When the cranking is stopped, a pressure peak is caused due to the rush of intake air rebounding off the piston, and the closing of the throttle butterfly.
7	Once all cranking has stopped, the waveform returns to atmospheric pressure.

Petrol intake manifold pressure idle speed

With the pressure transducer still connected to the intake manifold, the engine can now be started and the waveform analysed at idle.

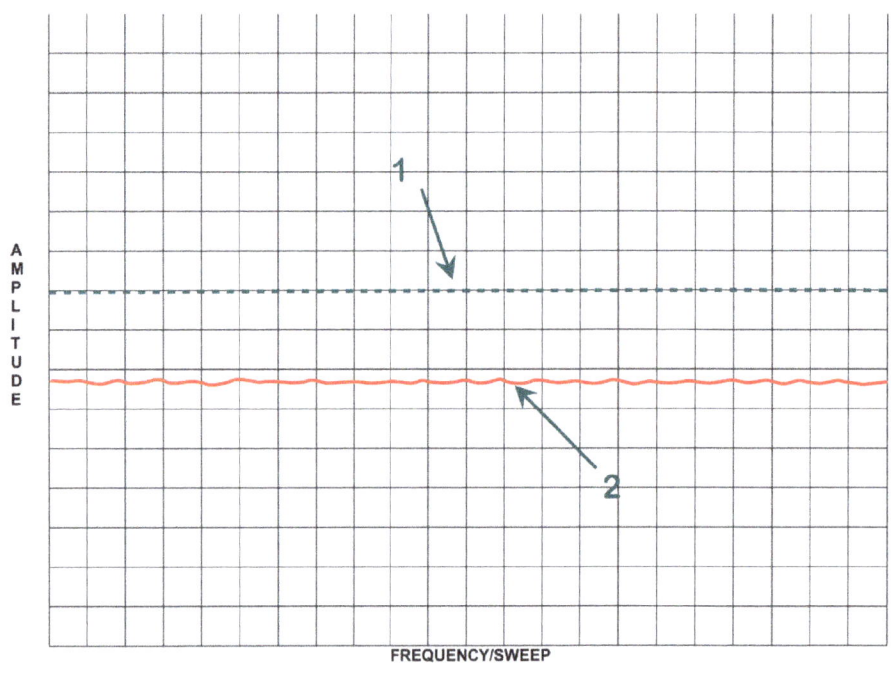

Figure 7.3 Intake manifold pressure while idling

Automotive Pressure Testing & Waveform Analysis

Table 7.3 Waveform analysis Intake manifold pressure (idling)

Waveform component	Description
1	The cursor line shown at this point represents atmospheric pressure.
2	With no zoom applied to the waveform image, it shows a steady ripple caused by the intake strokes of the pistons.
	As with cranking, the waveform should be relatively even, indicating a similar mechanical efficiency between all cylinders.

Petrol intake manifold pressure idle speed (zoomed)

In order to accurately assess the waveform produced by the pressure transducer from a petrol engine manifold while idling, it may be necessary to zoom in on the image if this function is available with your equipment. When zoomed, the waveform shows a ripple created by the opening and closing of the intake and exhaust valves, and the pattern should be checked for unusual or irregular pulsations that may indicate a fault. The next image represents a four-cylinder engine.

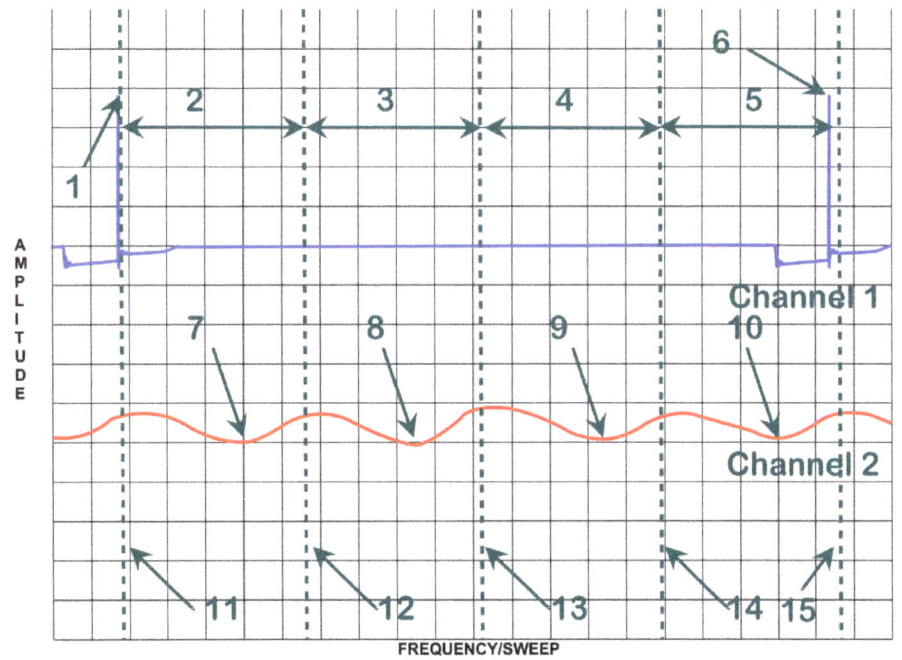
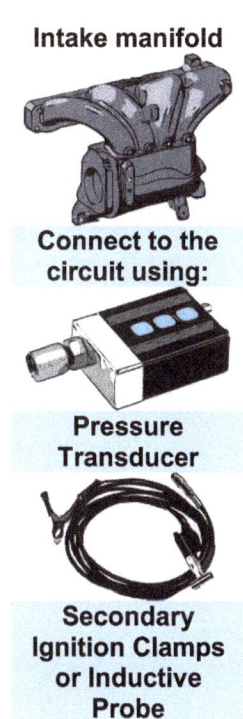

Figure 7.4 Intake manifold pressure while idling (zoomed)

Automotive Pressure Testing & Waveform Analysis

Table 7.4 Waveform analysis Intake manifold pressure (zoomed)

Waveform component	Description
1	Channel 1. A channel has been selected to show ignition from cylinder number one as a comparison against the pulsations from the zoomed intake manifold wave. This shows the firing voltage for number one cylinder and helps determine the start of the power stroke for that piston.
2	This is the power stroke period of cylinder number one.
3	This is the exhaust stroke period of cylinder number one.
4	This is the intake stroke period of cylinder number one.
5	This is the compression stroke period of cylinder number one.
6	Channel 1. This shows the firing voltage for number one cylinder, and helps determine the start of the power stroke for the next cycle of operations.
7	Channel 2. The dip on the waveform at this point represents the intake stroke of cylinder number four (on a 4-cylinder engine with the firing order 1342).
8	Channel 2. The dip on the waveform at this point represents the intake stroke of cylinder number two (on a 4-cylinder engine with the firing order 1342).
9	Channel 2. The dip on the waveform at this point represents the intake stroke of cylinder number one (on a 4-cylinder engine with the firing order 1342).
10	Channel 2. The dip on the waveform at this point represents the intake stroke of cylinder number three (on a 4-cylinder engine with the firing order 1342).
11	The cursor line at this point represents Top Dead Centre TDC of the pistons one and four on a 4-cylinder engine with a **flat-plane crankshaft**.
12	The cursor line at this point represents Bottom Dead Centre BDC of the pistons one and four on a 4-cylinder engine with a flat-plane crankshaft.
13	The cursor line at this point represents Top Dead Centre TDC of the pistons one and four on a 4-cylinder engine with a flat-plane crankshaft.
14	The cursor line at this point represents Bottom Dead Centre BDC of the pistons one and four on a 4-cylinder engine with a flat-plane crankshaft.
15	The cursor line at this point represents Top Dead Centre TDC of the pistons one and four on a 4-cylinder engine with a flat-plane crankshaft.

Flat-plane crankshaft – a standard crankshaft layout, with the crank pins set radially at an interval of 180º. This means that two companion pistons will go up and down together at the same time.

Automotive Pressure Testing & Waveform Analysis

Petrol intake manifold pressure snap throttle opening

With the pressure transducer still connected to the intake manifold and the engine running, the throttle can be rapidly opened and closed, with the response checked using the waveform. If a relatively slow time scale/sweep is selected, this will give you the opportunity to evaluate the intake manifold pressure under a number of different operating conditions:
- Idle
- Acceleration
- Wide open throttle (WOT)
- Engine off

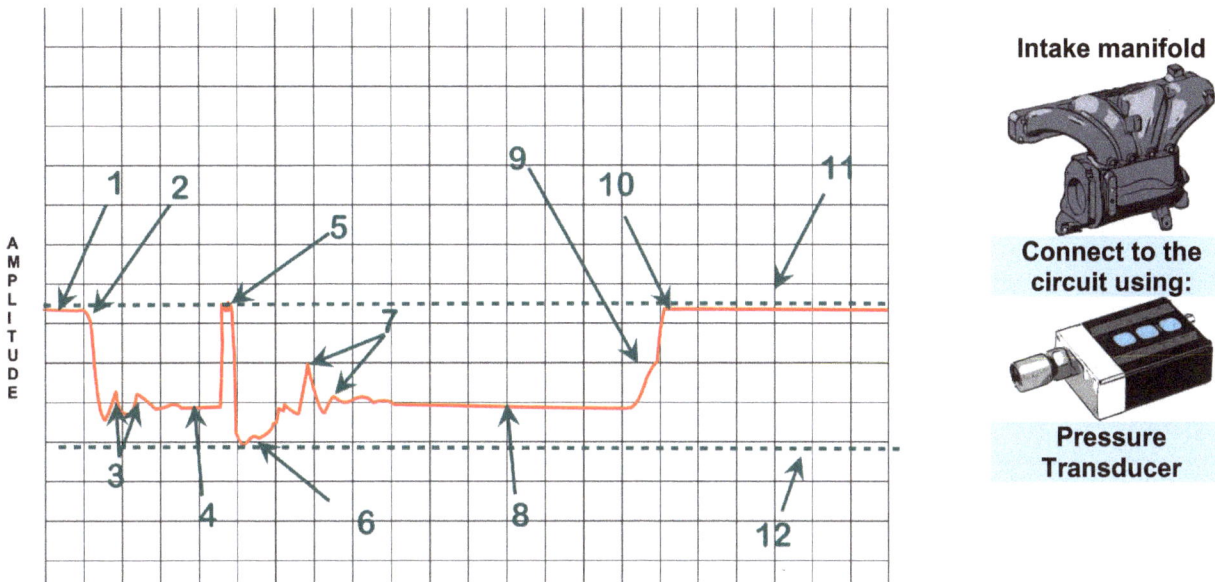

Figure 7.5 Intake manifold pressure with snap open throttle

Table 7.5 Waveform analysis Intake manifold pressure (snap open throttle)

Waveform component	Description
1	This shows the waveform with the engine switched off and will be the same as atmospheric pressure.
2	When the engine is cranked, and starts, the waveform will fall rapidly, representing the induction vacuum created by the running engine.
3	The small humps shown here at this point represent the engine trying to stabilise. This will vary from manufacturer to manufacturer, depending on various mechanical or electrical loads that are placed on the engine from components such as the alternator, power assisted steering and air conditioning for example.
4	Once the loads placed on the engine have been overcome by the idle control circuit, the vacuum in the manifold should show a fairly steady trace.

Automotive Pressure Testing & Waveform Analysis

Table 7.5 Waveform analysis Intake manifold pressure (snap open throttle)

5	With the engine at its correct operating temperature and the idle steady, the throttle can be quickly snapped open fully and released. The waveform should jump to atmospheric pressure and the rapidly fall back to vacuum. The rise and fall should be shown as a relatively sharp, but even curve.
6	As the throttle is snapped closed, the vacuum trace will momentarily dip below the idle value. This is due to the throttle butterfly restricting the airflow, and the mechanical intake energy caused by the descending pistons on their intake strokes raising the depression in the manifold. The larger the dip at this point, the greater the mechanical efficiency of the engine. (A small dip here could also indicate an intake manifold air leak).
7	As the engine returns to idle, there will once again be some disruption caused as the mechanical and electrical loads are sorted out by the idle control circuit; similar to that shown at position 3 on the diagram.
8	Once the loads placed on the engine have been overcome by the idle control circuit, the vacuum in the manifold should show a fairly steady trace.
9	As the engine is switched off, the vacuum in the manifold should rapidly return to atmospheric. A small disruption in the waveform may sometimes be shown as the crankshaft comes to a complete stop.
10	The vacuum with the engine stopped is now at atmospheric.
11	The cursor line here is shown to represent atmospheric pressure.
12	The cursor line here is shown to represent the maximum vacuum drawn during the test.

Exhaust pulses (cranking)

With an adapter attached to the pressure transducer, the oscilloscope may also be connected to the vehicle at the exhaust tail pipe, and the pulsations shown on the display can then be used to perform a mechanical diagnosis of engine operation.

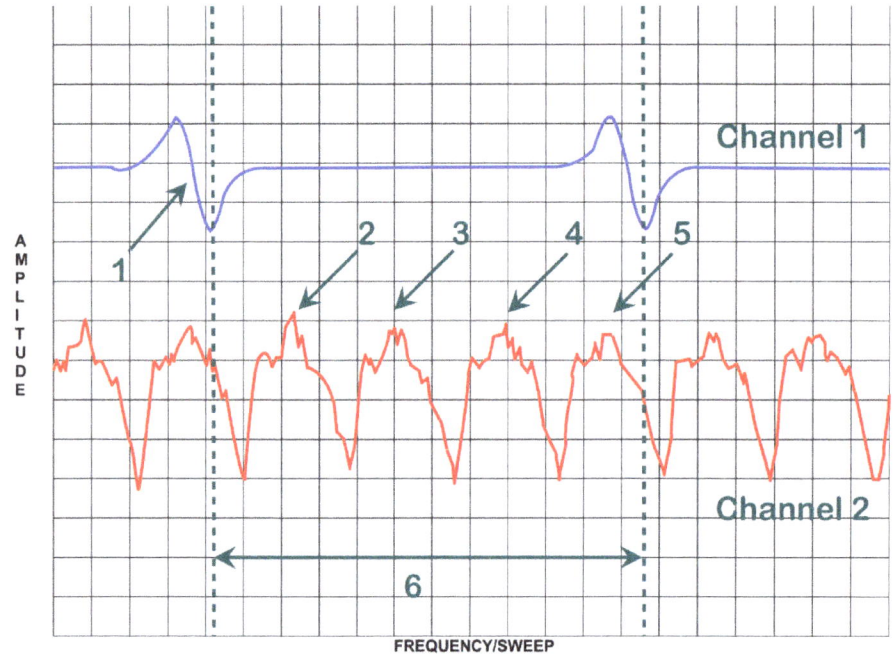

Figure 7.6 Exhaust tailpipe pressure with engine cranking

Automotive Pressure Testing & Waveform Analysis

Table 7.6 Waveform analysis Exhaust pulsations (cranking)

Waveform component	Description
1	Channel 1. This shows a channel connected to the camshaft position sensor as a reference signal. The pulse here will indicate the intake stroke of cylinder number one.
2	Channel 2. This peak represents the exhaust pulsation of cylinder 3, on a four-cylinder engine with the firing order 1342. The peak should be relatively even with the other cylinders displayed. A lower than expected peak at this point could indicate a poor mechanical efficiency from that cylinder.
3	Channel 2. This peak represents the exhaust pulsation of cylinder 4, on a four-cylinder engine with the firing order 1342. The peak should be relatively even with the other cylinders displayed. A lower than expected peak at this point could indicate a poor mechanical efficiency from that cylinder.
4	Channel 2. This peak represents the exhaust pulsation of cylinder 2, on a four-cylinder engine with the firing order 1342. The peak should be relatively even with the other cylinders displayed. A lower than expected peak at this point could indicate a poor mechanical efficiency from that cylinder.
5	Channel 2. This peak represents the exhaust pulsation of cylinder 1, on a four-cylinder engine with the firing order 1342. The peak should be relatively even with the other cylinders displayed. A lower than expected peak at this point could indicate a poor mechanical efficiency from that cylinder.
6	The cursor lines shown on the display are laid out to represent one complete revolution of the crankshaft.

Exhaust pulses (running)

With the engine running, the pulsations created by the exhaust pressure at the tailpipe will increase in amplitude and frequency. Due to the speed of operation, the pulsations shown on the waveform will appear much smoother than those when the engine is cranking.

Depending on where you connect the pressure transducer to the exhaust system, the amplitude shown on the waveform can give you an indication of a blockage or restriction.

Automotive Pressure Testing & Waveform Analysis

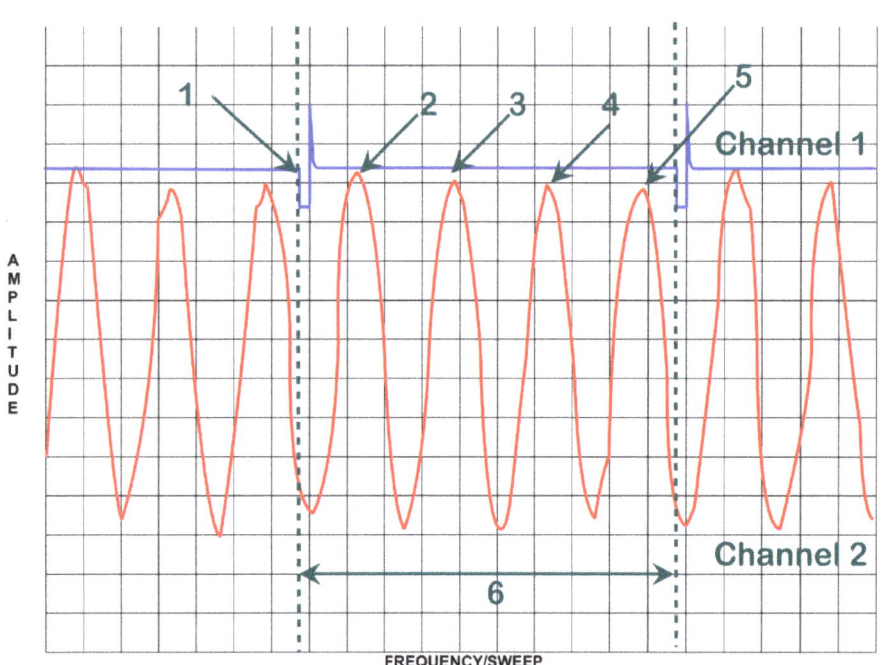
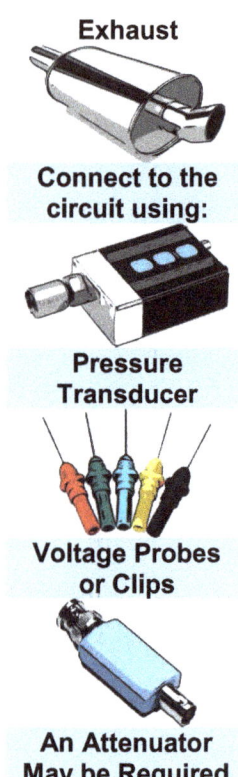

Figure 7.7 Exhaust tailpipe pressure with engine running

Table 7.7 Waveform analysis Exhaust pulsations (running)

Waveform component	Description
1	Channel 1. This shows a channel connected to number one fuel injector as a reference signal. The pulse here will give an indication of the intake stroke for cylinder number one.
2	Channel 2. This peak represents the exhaust pulsation of cylinder 3, on a four-cylinder engine with the firing order 1342. The peak should be relatively even with the other cylinders displayed. A lower than expected peak at this point could indicate a poor mechanical efficiency from that cylinder.
3	Channel 2. This peak represents the exhaust pulsation of cylinder 4, on a four-cylinder engine with the firing order 1342. The peak should be relatively even with the other cylinders displayed. A lower than expected peak at this point could indicate a poor mechanical efficiency from that cylinder.

Automotive Pressure Testing & Waveform Analysis

Table 7.7 Waveform analysis Exhaust pulsations (running)

4	Channel 2. This peak represents the exhaust pulsation of cylinder 2, on a four-cylinder engine with the firing order 1342. The peak should be relatively even with the other cylinders displayed. A lower than expected peak at this point could indicate a poor mechanical efficiency from that cylinder.
5	Channel 2. This peak represents the exhaust pulsation of cylinder 1, on a four-cylinder engine with the firing order 1342. The peak should be relatively even with the other cylinders displayed. A lower than expected peak at this point could indicate a poor mechanical efficiency from that cylinder.
6	The cursor lines shown on the display are laid out to represent one complete revolution of the crankshaft.

In-cylinder compressions (cranking)

With a dedicated adapter attached to the pressure transducer, the oscilloscope may also be connected to a petrol engine at the spark plug fitting, and the pulsations shown on the display can then be used to perform a compression diagnosis of a single cylinder during engine operation.
To test the cranking compression, the engine must be isolated so that it doesn't start.

Due to engine design, the exact shape of waveform will vary from manufacturer to manufacturer.

A pressure transducer adapter should only be used/connected to recommended systems that operate within the manufacturing tolerance of the equipment. Exceeding the measurable/recommended pressures can lead to personal injury, vehicle and equipment damage.

It is also very important to remember that connecting a pressure transducer to a cylinder for in-cylinder compression testing will cause that cylinder not to operate in the normal manner. The misfire created by removing the spark plug and inserting the adapter could lead to catalytic converter damage. The **ignition and fuel injection MUST be isolated** during testing to reduce the possibility of damage or injury.

It is also recommended that any in-cylinder compression testing is conducted for the shortest possible amount of time required to gather diagnostic information.

Automotive Pressure Testing & Waveform Analysis

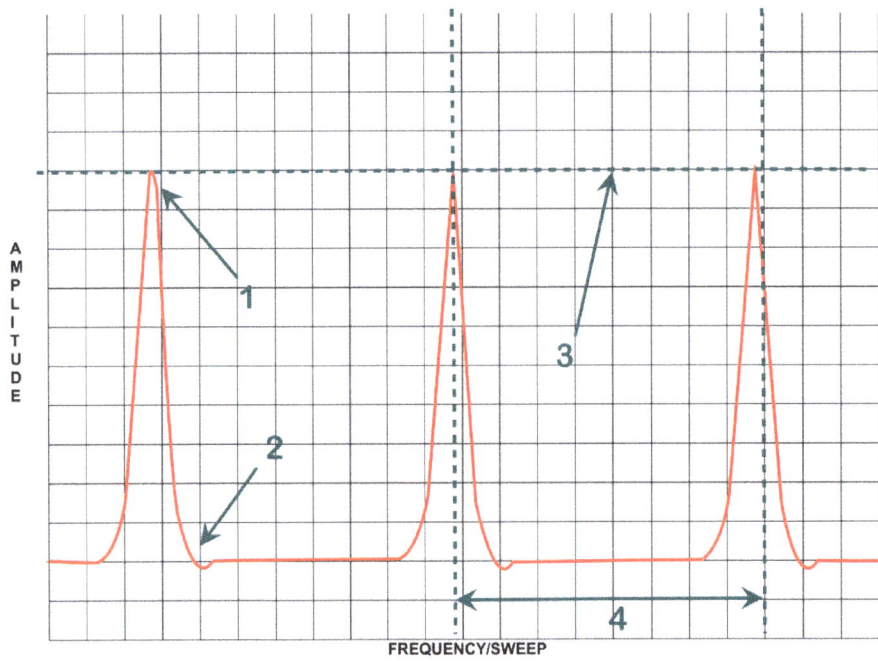
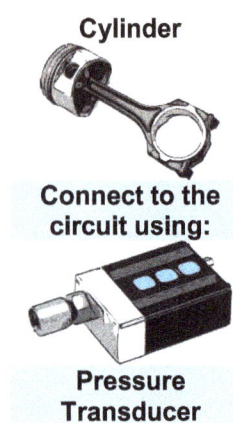

Figure 7.8 In-cylinder compression pressures with the engine cranking

Table 7.8 Waveform analysis In-cylinder pressure (cranking)

Waveform component	Description
1	This shows the compression pressure of the cylinder being tested while the engine is being cranked. The amplitude will give an indication of engine cylinder mechanical efficiency, but will also depend on the cranking speed of the starter motor. In order to obtain the most efficient intake of air during cranking, the throttle butterfly should be held wide open.
2	This small dip shown, at the equivalent end of the power stroke, is caused by the opening of the exhaust valve as any final pressure in the cylinder drops.
3	This cursor line represents the highest peak compression measured during cranking, and should be similar to the values expected during conventional compression testing.
4	The two cursors shown here represent one complete revolution of the crankshaft.

Tip: If a compression peak were divided vertically in half, it should show a very symmetrical rise and fall in amplitude.
This can be used as an indication of the cylinders' ability to seal effectively at the valves and piston rings.

Automotive Pressure Testing & Waveform Analysis

In-cylinder compression (running)

With the pressure transducer attached to the adapter and set up to measure in-cylinder compression, the engine can now be started and the waveform analysed. Remember that with the spark plug removed, the engine will be running with one less cylinder. This will have an effect on the smoothness of operation and low speed idle may not be achievable. To get a suitable waveform for diagnosis, it will be necessary to raise engine speed.

Running the engine with a spark plug removed will normally cause a diagnostic trouble code to be generated, and the malfunction indicator lamp (MIL) may illuminate. It is important to ensure that all diagnostic trouble codes have been cleared after any work has been conducted.

A pressure transducer adapter should only be used/connected to recommended systems that operate within the manufacturing tolerance of the equipment. Exceeding the measurable/recommended pressures can lead to personal injury, vehicle and equipment damage.

It is also very important to remember that connecting a pressure transducer to a cylinder for in-cylinder compression testing will cause that cylinder not to operate in the normal manner. The misfire created by removing the spark plug and inserting the adapter could lead to catalytic converter damage. The **fuel injection MUST be isolated** during testing to reduce the possibility of damage or injury.

It is also recommended that any in-cylinder compression testing is conducted for the shortest possible amount of time required to gather diagnostic information.

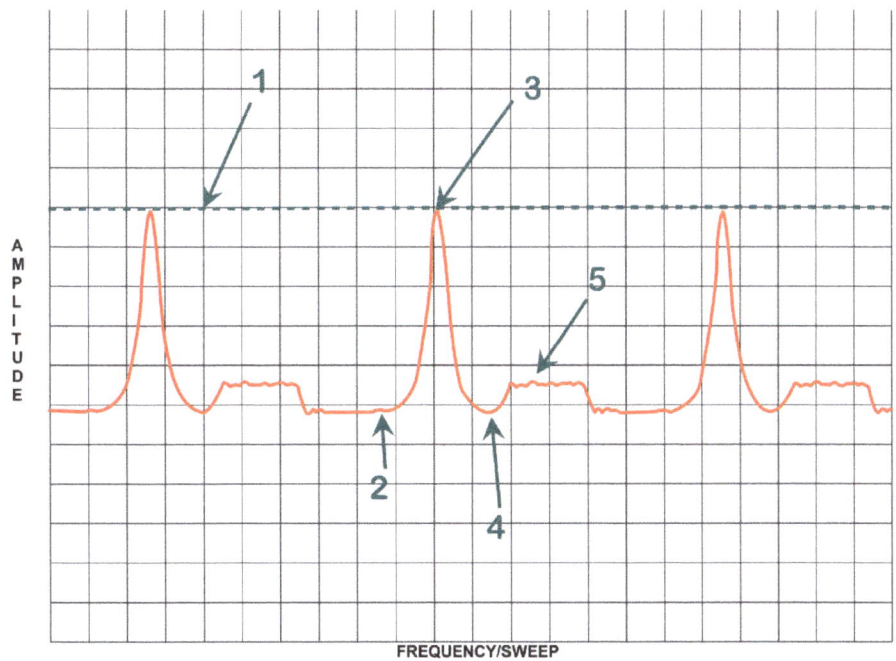
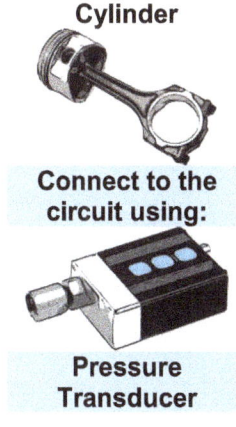

Figure 7.9 In-cylinder compression pressures with the engine running

Automotive Pressure Testing & Waveform Analysis

Table 7.9 Waveform analysis In-cylinder pressure (running)

Waveform component	Description
1	The cursor line added to this screen shows the maximum amplitude reached by the compression on the cylinder being tested.
2	The small disruptions to the waveform are caused by the closing of the inlet valve, just before the beginning of the compression stoke.
3	This point shows the peak compression reached by the cylinder under test. If the waveform was divided vertically at this point, a symmetrical rise and fall of compression shows a good sealing of the combustion chamber.
4	The dip in pressure here is caused by the opening of the exhaust valve, just before the piston has reached the end of its power stroke.
5	A small pressure rise can be seen here as the piston moves back up the cylinder bore and exhaust gas is pushed out of the exhaust valve.

Crankcase pressure (cranking)

By connecting the pressure transducer to the engine's crankcase ventilation (breather) system, the oscilloscope can be used to perform a diagnosis of any leakage of compression pressure past the pistons during engine operation. In order to work out which cylinder is creating which crankcase pulsation, it will be necessary to have a reference signal set up on a second channel.

To test the cranking pressures, the engine will have to be isolated so that it doesn't start.

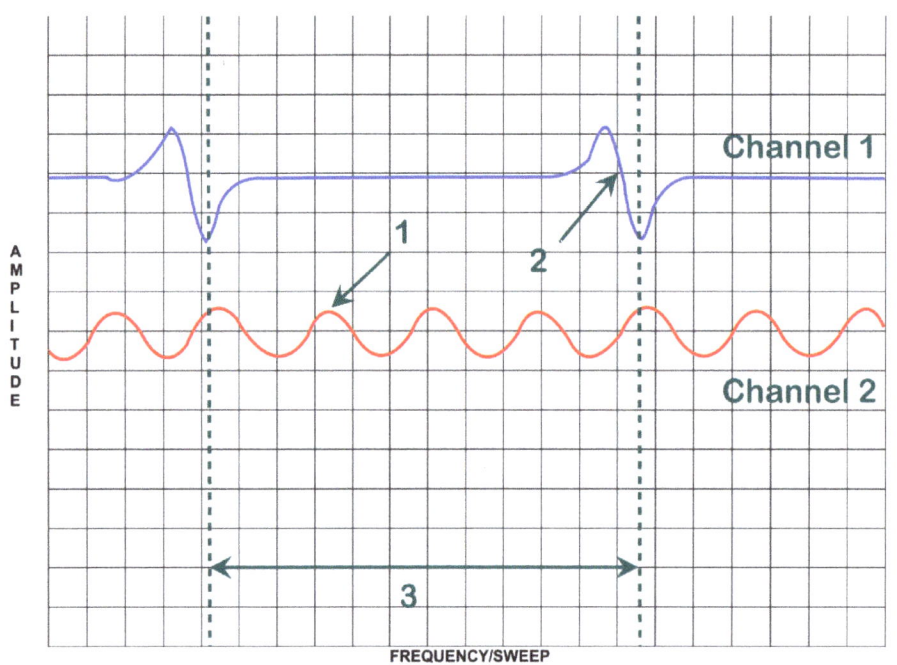

Figure 7.10 Crankcase pressures with the engine cranking

Automotive Pressure Testing & Waveform Analysis

Table 7.10 Waveform analysis Crankcase pressure (cranking)

Waveform component	Description
1	Channel 2. With a channel set up as a reference signal using the camshaft position sensor, this pulse should indicate the compression stroke of cylinder number one. All pulsations should be relatively even. A higher than expected amplitude on one cylinder may indicate 'blow-by' pressure leaking past piston rings. The following pulsations will indicate the compression strokes of the other cylinders in the firing order of the engine.
2	Channel 1. This shows the induced wave created by the camshaft sensor so that it can be compared as a reference signal against the crankcase pressure pulsations. This will indicate the intake stroke of cylinder number one, so the following crankcase pulsation will be the compression stroke.
3	The cursors shown on the display are used to represent two complete revolutions of the crankshaft.

Crankcase pressure (running)

With the ignition and fuel injection system reconnected, the engine can now be started and crankcase pressures measured while the engine is running. A reference signal will once again be required, and the amplitude and frequency of the crankcase pressures will increase due to the speed of the running engine.

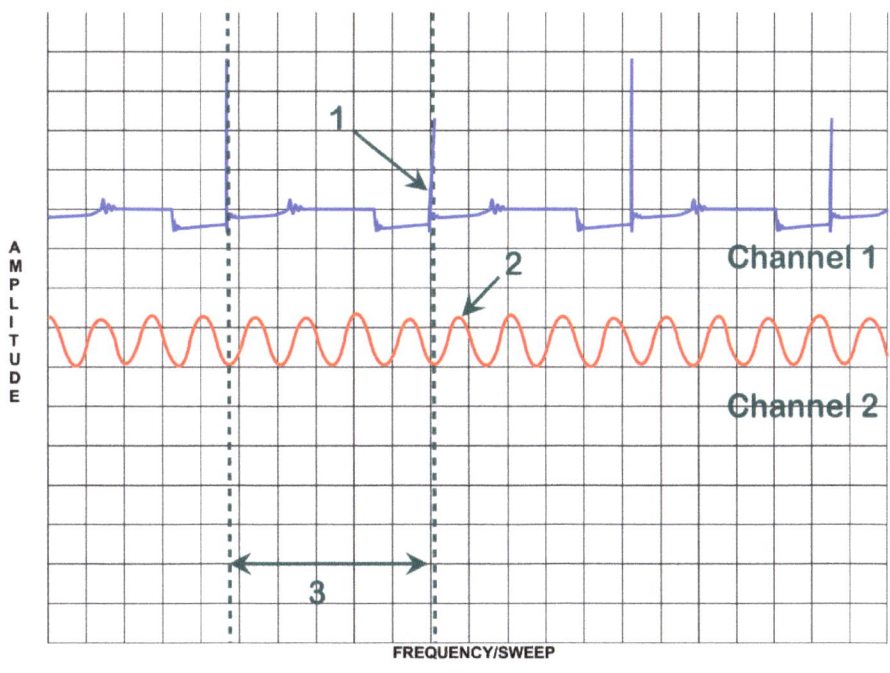

Figure 7.11 Crankcase pressures with the engine running

Automotive Pressure Testing & Waveform Analysis

Table 7.11 Waveform analysis Crankcase pressure (running)

Waveform component	Description
1	Channel 1. With a channel set up as a reference signal, using the secondary ignition waveform of cylinder number one, this point will show the firing impulse of the spark plug at the end of the compression stroke.
2	Channel 2. The peaks shown here are a combination of both compression and power strokes. Using channel 1 as a reference, this is the pulsation of cylinder number one; all pulsations should be relatively even. A higher than expected amplitude on one cylinder may indicate 'blow-by' pressure leaking past piston rings.
3	The cursors shown on the display are used to represent one complete revolution of the crankshaft of a wasted spark engine.

Fuel pressure regulator verses injection (petrol)

A petrol fuel pressure regulator can also be measured using a pressure transducer device attached to the oscilloscope. The transducer should be connected to the vacuum port of the pressure regulator. With a fuel injector pattern set-up on a second channel as a reference, pressure variations in the fuel rail caused by the opening and closing of the injectors can be seen.

The actual pattern shown on the oscilloscope will depend on the design of the fuel injection system and whether the injectors open as simultaneous, grouped or sequential.
Take care when analysing patterns, and if unsure, compare with another similar vehicle that has no known operating issues.

A pressure transducer adapter should only be used/connected to recommended systems that operate within the manufacturing tolerance of the equipment. Exceeding the measurable/recommended pressures can lead to personal injury, vehicle and equipment damage.

Automotive Pressure Testing & Waveform Analysis

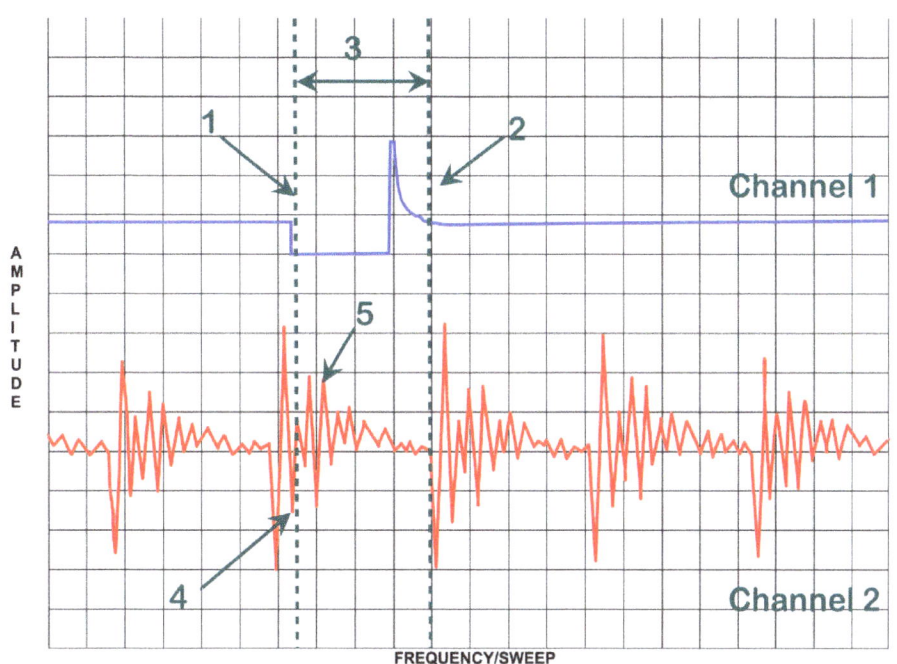
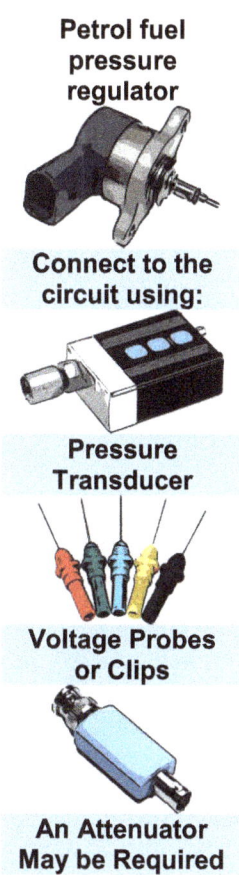

Figure 7.12 Fuel pressures verses injector operation

Table 7.12 Waveform analysis petrol fuel pressure regulator

Waveform component	Description
1	Channel 1. With channel 1 set-up to represent the waveform created by injector number one, this cursor line has been set to show the opening of the injector pintle, just after the ECU has energised the internal solenoid coil.
2	Channel 1. This cursor line has been set to show the closing of the injector pintle, just after the ECU has switched off the internal solenoid coil. As the spring closes the pintle, the armature movement causes a disruption to the magnetic field and creates a small hump in the waveform of the back EMF (see injector waveform analysis).
3	This represents the injector pintle open period.
4	Channel 2. When the injector opens, there is a sudden drop in fuel pressure as indicated here by the transducer measurement.
5	Channel 2. After the initial pressure drop, caused by the opening of the injector has passed, a good fuel pump should quickly return pressure to the regulator as the waveform pattern shows.

Automotive Pressure Testing
& Waveform Analysis

If the drop in fuel pressure on any one of the openings is less than another, this could indicate a blocked or inoperative injector.

Common Acronyms/Abbreviations

A - Amperes
A/C - Air Conditioning
A/F - Air Fuel Ratio
A/T - Automatic Transmission
AAC - Auxiliary Air Control Valve
AAT - Ambient Air Temperature
ABD - Automatic Brake Differential
ABS - Antilock Brake System
ABV - Air Bypass Valve
AC - Alternating Current
ACC - Automatic Climate Control
ACC - Air Conditioning Clutch
ACR - Air Conditioning Relay
ACR4 - Air Conditioning Refrigerant, Recovery, Recycling, Recharging
ACV - Air Control Valve
ADU - Analogue-Digital Unit
AEV - All Electric Vehicle
AFC - Air Flow Control
AFL - Advanced Front Lighting System
AFM - Air Flow Meter
AFR - Air Fuel Ratio
AFS - Air Flow Sensor
AGM - Absorbed Glass Matt
AGS - Adaptive Gearbox Control (network acronym)
Ah - Amp Hours
AIR - Secondary Air Injection System
AIS - Automatic Idle Speed
ALC - Automatic Level Control
AM - Amplitude Modulation
API - American Petroleum Institute
APS - Atmospheric Pressure Sensor
ARC - Automatic Ride Control
ARS - Automatic Restraint System
ASARC - Air Suspension Automatic Ride Control
ATC - Automatic Temperature Control
ATDC - After Top Dead Centre
ATF - Automatic Transmission Fluid
ATS - Air Temperature Sensor
AVO - Amps Volts Ohms
AWD - All Wheel Drive
AWG - American Wire Gage
AYC - Active Yaw Control
B/MAP - Barometric/Manifold Absolute Pressure
BARO - Barometric Pressure
BCM - Body Control Module
BCM - Battery Control Module
BDC - Bottom Dead Centre
BEV - Battery Electric Vehicle
BHP - Brake Horsepower
BOB - Breakout Box

BP - Barometric Pressure
BPP - Brake Pedal Position Switch
BTDC - Before Top Dead Centre
BTS - Battery Temperature Sensor
Btu - British thermal unit
BUS N - Bus Negative
BUS P - Bus Positive
C - Celsius
CA - Cranking Amps
CAN - Controller Area Network
CANP - EVAP Canister Purge Solenoid
CAS - Crank Angle Sensor
CBW - Clutch by Wire
CC - Catalytic Converter
CC - Climate Control
CC - Cruise Control
CC - Cubic Centimetres
CCA - Cold Cranking Amps
CD - Compact Disc
CDC - Compact Disc Changer (network acronym)
CDI - Capacitor Discharge Ignition
CFC - Chlorofluorocarbons
CFI - Continuous Fuel Injection
CI - Compression Ignition
CKP - Crankshaft Position Sensor
CL - Closed Loop
CLC - Converter Lockup Clutch
CLV - Calculated Load Value
CMP - Camshaft Position Sensor
CNG - Compressed Natural Gas
CO - Carbon Monoxide
CO2 - Carbon Dioxide
COC - Conventional Oxidation Catalyst
COP - Coil on Plug Electronic Ignition
COSHH - Control of Substances Hazardous to Health
CP - Crankshaft Position Sensor
CP - Canister Purge (GM)
CPP - Clutch Pedal Position
CPU - Central Processing Unit
CRC - Cyclic Redundancy Check
CRD - Common Rail Diesel
CRS - Common Rail System
CRT - Cathode Ray Tube
CTP - Closed Throttle Position
CTS - Coolant Temperature Sensor
CV - Constant Velocity
CVT - Continuously Variable Transmission
DBW - Drive by Wire
DC - Duty Cycle
DC - Direct Current
DCS - Dual Clutch System

Common Acronyms/Abbreviations

DI - Distributor Ignition (System)
DI - Direct Ignition
DIS - Direct Ignition (Waste Spark)
DIS - Distributor-less Ignition System
DME - Engine Management (network acronym)
DMF - Dual Mass Flywheel
DMM - Digital Multimeter
DLC - Data Link Connector (OBD)
DOHC - Dual Overhead Cam
DPF - Diesel Particulate Filter
DRL - Daytime Running Lights
DSC - Dynamic Stability Control (network acronym)
DSO - Digital Storage Oscilloscope
DTC - Diagnostic Trouble Code
DVD - Digitally Versatile Disc
EAIR - Electronic Secondary Air Injection
EBCM - Electronic Brake Control Module
EBP - Exhaust Back Pressure
EBD - Electronic Brake Force Distribution
ECC - Electronic Climate Control
ECM - Engine/Electronic Control Module
ECS - Emission Control System
ECT - Engine Coolant Temperature
ECU - Electronic Control Unit
EDC - Electronic Diesel Control
EECS - Evaporative Emission Control System
EEGR - Electronic EGR (Solenoid)
EEPROM - Electronically Erasable Programmable Read Only Memory
EFI - Electronic Fuel Injection
EFT - Engine Fuel Temperature
EGO - Exhaust Gas Oxygen Sensor
EGR - Exhaust Gas Recirculation
EGRT - Exhaust Gas Recirculation Temperature
EMF - Electromotive Force (voltage)
EMI - Electromagnetic Interference
EML - Engine Management Light
EOBD - European On Board Diagnostics
EOP - Engine Oil Pressure
EOT - Engine Oil Temperature
EPA - Environmental Protection Act
EPB - Electronic Parking Brake
EPROM - Erasable Programmable Read Only Memory
EPS - Electronic Power Assisted Steering
ESP - Electronic Stability Programme
ESS - Engine Start-Stop
EVAP - Evaporative Emissions System
EVAP CP - Evaporative Canister Purge
EWS - Immobiliser (network acronym)
FM - Frequency Modulation
FOT - Fixed Orifice Tube

FSD - Full Scale Deflection
FT - Fuel Trim
FWD - Front Wheel Drive
GDI - Gasoline Direct Injection
GND - Electrical Ground Connection
GPS - Global Positioning System
GWP - Global Warming Potential
H - Hydrogen
HASAWA - Health and Safety at Work Act
H2O - Water
HC - Hydrocarbons
HCA - Hot Cranking Amps
HDI - High Pressure Direct Injection
HEGO - Heated Exhaust Gas Oxygen Sensor
HFC - Hydrogen Fuel Cell
HFC - Hydro-fluoro Carbon
HFO - Hydro-fluoro Olefin
Hg - Mercury
HICE - Hydrogen Internal Combustion Engine
HID - High Intensity Discharge (lighting)
HO2S - Heated Oxygen Sensor
hp - Horsepower
HSE - Health and Safety Executive
HT - High Tension
HUD - Head up Display
HVAC - Heating Ventilation and Air Conditioning
Hz - Hertz
I/O - Input / Output
IA - Intake Air
IAC - Idle Air Control (motor or solenoid)
IAT - Intake Air Temperature
IC - Integrated Circuit
IC - Ignition Control
ICE - In Car Entertainment
ICE - Internal Combustion Engine
ICM - Ignition Control Module
IFS - Inertia Fuel Switch
IGBT - Insulated Gate Bipolar Transistor
IGN - Ignition
IGN ADV - Ignition Advance
IGN GND - Ignition Ground
IHKA – Climate Control (network acronym)
IPR - Injector Pressure Regulator
ISC - Idle Speed Control
ISO - International Standard of Organisation
KAM - Keep Alive Memory
Kg/cm2 - Kilograms/ Cubic Centimetres
KHz - Kilohertz
Km - Kilometres
Kombi – Instrument Cluster (network acronym)
KPA - Kilopascal
KPI - Kingpin Inclination

Common Acronyms/Abbreviations

KS - Knock Sensor
KWP - Keyword Protocol
l - Litres
LCD - Liquid Crystal Display
LED - Light Emitting Diode
LHD - Left Hand Drive
Li-ion - Lithium ion
LOOP - Engine Operating Loop Status
LOS - Limited Operating Strategy
LPG - Liquefied Petroleum Gas
LSD - Limited Slip Differential
LTFT - Long Term Fuel Trim
LWB - Long Wheel Base
LWR - Vertical Headlight Control (network acronym)
M/T - Manual Transmission
MAC - Mobile Air Conditioning
MAF - Mass Air Flow Sensor
MAP - Manifold Absolute Pressure Sensor
MAT - Manifold Air Temperature
MCM - Motor Control Module
MEF - Methane Equivalency Factor
MF - Maintenance Free
MFI - Multiport Fuel Injection
MIL - Malfunction Indicator Lamp
MPG - Miles per Gallon
MPH - Miles per Hour
MRS - Multiple Restraint System (network acronym)
mS or ms - Millisecond
mV or mv - Milivolt
N - Nitrogen
NCAPS - Non-Contact Angular Position Sensor
NCRPS - Non-Contact Rotary Position Sensor
NGV - Natural Gas Vehicles
Ni-MH - Nickel Metal Hydride
Nm - Newton Meters
NOx - Oxides of Nitrogen
NPN - Negative Positive Negative
NTC - Negative Temperature Coefficient
O2 - Oxygen
OBD I - On Board Diagnostics Version I
OBD II - On Board Diagnostics Version II
OC - Oxidation Catalytic Converter
OD - Overdrive
OD - Outside Diameter
ODP - Ozone Depletion Potential
OE - Original Equipment
OEM - Original Equipment Manufacturer
OFN - Oxygen Free Nitrogen
OHC - Overhead Cam Engine
OHV - Overhead Valve
OL - Open Loop

OS - Oxygen Sensor
P/N - Part Number
PAG - Polyalkylene Glycol
PAIR - Pulsed Secondary Air Injection
PATS - Passive Anti-Theft System
PCB - Printed Circuit Board
PCM - Powertrain Control Module
PCV - Positive Crankcase Ventilation
Pd - Potential Difference (volts)
PD - Pumpe-Düse
PEF - Propane Equivalency Factor
PEM - Proton Exchange Membrane
PFI - Port Fuel Injection
PGM-FI - Programmed Gas Management Fuel Injection
PID - Parameter Identification Location
PKE - Passive Keyless Entry
PNP - Positive Negative Positive
POF - Plastic Optical Fibre
POT - Potentiometer
PPE - Personal Protective Equipment
PPM - Parts Per Million
PPS - Accelerator Pedal Position Sensor
PROM - Programmable Read-Only Memory
PSI - Pounds per Square Inch
PTC - Positive Temperature Coefficient Resistor
PTO - Power Take Off (4WD Option)
PUWER - Provision and Use of Work Equipment Regulations
PWM - Pulse Width Modulation
RAM - Random Access Memory
RBS - Regenerative Braking system
RCM - Reserve Capacity Minutes
RDS - Radio Data System
RDW - Tyre Pressure Monitoring (network acronym)
REF - Reference
RFI - Radio Frequency Interference
RHD - Right Hand Drive
RIDDOR - Reporting of Injuries Diseases and Dangerous Occurrence Regulations
RKE - Remote Keyless Entry
RMS - Recovery Management Station
ROM - Read Only Memory
RON - Research Octane Number
RTV - Room Temperature Vulcanizing
RWD - Rear Wheel Drive
SAE - Society of Automotive Engineers (Viscosity Grade)
SAI - Swivel Axis Inclination
SCR - Selective Catalytic Regeneration
SCS - Sick Car Syndrome

Common Acronyms/Abbreviations

SFI - Sequential Fuel Injection
SI - Spark Ignition
SIPS - Side Impact Protections System
SOC - State of Charge
SOHC - Single Overhead Cam
SPFI - Single Point Fuel Injection (throttle body)
SRI - Service Reminder Indicator
SRS - Supplementary Restraint System (air bag)
SRT - System Readiness Test
STFT - Short-Term Fuel Trim
SWB - Short Wheel Base
SWL - Safe Working Load
SZM - Central Switch Module (network acronym)
TAC - Throttle Actuator Control
TACH - Tachometer
TBI - Throttle Body Injection
TC - Turbocharger
TCC - Torque Converter Clutch
TCM - Transmission Control Module
TCS - Traction Control System
TD - Turbo Diesel
TDC - Top Dead Centre
TDI - Turbo Direct Injection
TOOT - Toe Out On Turns
TP - Throttle Position
TPM - Tyre Pressure Monitor
TPP - Throttle Position Potentiometer
TPS - Throttle Position Sensor
TSB - Technical Service Bulletin
TV - Throttle Valve
TXV - Thermal Expansion Valve
UART - Universal Asynchronous Receiver-Transmitter
UJ - Universal Joint
USB - Universal Serial Bus
UV - Ultraviolet
V - Volts
VAC - Vacuum
VAF - Vane Airflow Meter
VDP - Variable Diameter Pulley
VDU - Visual Display Unit
VIN - Vehicle Identification Number
VPE - Vehicle Protection Equipment
VSS - Vehicle Speed Sensor
W/B - Wheelbase
WOT - Wide Open Throttle
WSS - Wheel Speed Sensor
YRS - Yaw Rate Sensor

Appendix
DIN Terminal Numbers

Ignition system

1	coil, distributor, low voltage
1a, 1b	distributor with two separate circuits
2	breaker points magneto ignition
4	coil, distributor, high voltage
4a, 4b	distributor with two separate circuits, high voltage
7	terminal on ballast resistor, to distributor
15	battery+ from ignition switch
15a	from ballast resistor to coil and starter motor

Preheat (Diesel engines)

15	preheat in
17	start
19	preheat (glow)

Starter

50	starter control

Battery

15	battery+ through ignition switch
30	from battery+ direct
30a	from 2nd battery and 12/24 V relay
31	return to battery- or direct to ground
31a	return to battery- 12/24 V relay
31b	return to battery- or ground through switch
31c	return to battery- 12/24 V relay

Electric motors

32	return
33	main terminal (swap of 32 and 33 is possible)
33a	limit
33b	field
33f	2. slow rpm
33g	3. slow rpm
33h	4. slow rpm
33L	rotation left
33R	rotation right

Turn indicators

49	flasher unit in
49a	flasher unit out, indicator switch in
49b	out 2. flasher circuit
49c	out 3. flasher circuit
C	1st flasher indicator light
C2	2nd flasher indicator light
C3	3rd flasher indicator light
L	indicator lights left
R	indicator lights right
L54	lights out, left
R54	lights out, right

AC generator (alternator)

51	DC at rectifiers
51e	as 51, with choke coil
59	AC out, rectifier in, light switch
59a	charge, rotor out
64	generator control light

Generator, voltage regulator

61	charge indicator (charge control light)
B+	battery +
B-	battery -
D+	dynamo +
D-	dynamo -
DF	dynamo field
DF1	dynamo field 1
DF2	dynamo field 2
U, V, W	AC three phase terminals

Lights

54	brake lights

Appendix
DIN Terminal Numbers

55	fog light	82z	1st in
56	spot light	82y	2nd in
56a	headlamp high beam and indicator light	83	multi position switch, in
56b	low beam	83a	out position 1
56d	signal flash	83b	out position 2
57	parking lights	**Relay**	
57a	parking lights	85	relay coil -
57L	parking lights left	86	relay coil +
57R	parking lights right	**Relay contacts**	
58	licence plate lights, instrument panel	87	common contact
58d	panel light dimmer	87a	normally closed contact

Window wiper/washer

		87b	normally open contact
53	wiper motor + in	88	common contact 2
53a	limit stop+	88a	normally closed contact 2
53b	limit stop field	88b	normally open contact 2
53c	washer pump	**Additional**	
53e	stop field	52	signal from trailer
53i	wiper motor with permanent magnet, third brush for high speed	54g	magnetic valves for trailer brakes
		75	radio, cigarette lighter
		77	door valves control

Acustic warning

71	beeper in
71a	beeper out, low
71b	beeper out, high
72	hazard lights switch
85c	hazard sound on

Switches

81	opener
81a	1 out
81b	2 out
82	lock in
82a	1st out
82b	2nd out

Index

ABS, 33, 39, 81, 82, 87, 91, 92, 93, 189
active sensor, 86
Actuators, 18, 32, 35
air temperature sensor, 89, 112, 113
Alternating current (AC), 8
Alternator, 33, 39, 75, 76, 77, 79
Amp clamp, 28
Amplitude, 7, 27
Amps, 8
Amps draw, 17
Analogue waveforms, 19
Arbitration, 152
Armature, 39
Attenuator, 26
axis, 22
Back EMF, 157
Back-probe, 26
bad earth, 16
Banana Plug, 23
barometric, 89, 113, 114
BNC, 23
breakout box, 18
camshaft sensor, 73, 89, 114, 115, 116, 117, 185
CAN bus, 155, 156, 158, 160, 161, 162
catalyst, 90, 91, 124, 125
catalytic, 90, 91, 120, 124, 125, 181, 183
Channels, 30
Circuits, 11
Closed loop, 117, 123
coil on plug, 25, 131, 146, 147, 148
common rail Diesel, 58
Commutator, 40
Conductors, 11
Consumer, 13
Continuity, 11

coolant temperature sensor, 110, 111, 112
Cooling fan, 38, 68
Corruption, 160
Crankcase pressure, 170, 184, 185
crankshaft sensor, 88, 89, 91, 108, 109, 110
Current, 9
Cursors, 30, 32
Dead short, 16
Delta, 31
Depression, 92
Digital waveforms, 19
DIN terminal numbers, 18
Direct current (DC), 8
distributor, 89, 133, 134, 193
Drive-by-wire, 92
duty cycle, 19
Duty cycle and PWM, 19
Dwell period, 136
EBD, 87, 92, 190
ECU, 18, 32
EGR, 33, 36, 45, 46, 102, 190
Electromotive force (EMF), 8, 9
electrons, 10
ESP, 40
EVAP, 33, 36, 41, 42, 189, 190
Exhaust pulses, 178, 179
Filters, 30, 32
Flat-plane crankshaft, 176
FlexRay, 150, 152, 165, 166, 168
frequency, 6, 31
fuel pressure regulator, 186
fuel pressure sensor, 90, 117, 118
fuel pump, 36, 46, 47, 187
Gateway, 157
Generators, 12

Index

glow plugs, 36, 43, 44

Hall effect, 26, 88, 89, 91, 95, 96, 97, 109, 110, 114, 115, 127, 133

Hertz (Hz)., 31

Hexadecimal, 160

high resistance, 16

High-tension HT, 135

IAC, 37

idle air control valve, 48

Ignition primary, 131, 135, 136

In-cylinder, 170, 181, 182, 183

Inductive, 25, 26, 28, 43, 44, 47, 54, 55, 57, 59, 60, 61, 62, 64, 71, 73, 74, 76, 78, 79, 86, 87, 89, 93, 94, 108, 116, 123, 137, 138, 139, 140, 142, 144, 146, 148, 175, 185

Inertia, 72, 100, 102

Infotainment, 160

injector, 37, 51, 52, 53, 54, 55, 56, 57, 58, 59, 60, 61, 62, 63, 64, 65, 118, 120, 180, 186, 187, 188

insulators, 11

intake manifold pressure, 170, 173, 174, 175, 176, 177

K-line, 167

knock sensor, 90, 118, 119, 120

Lambda sensor, 83, 90, 120, 121, 122, 123

LIN, 150, 152, 168, 169

Local Interconnect, 152, 168

Low-tension LT, 135

Magnetic flux, 93

Magnets, 12

manifold absolute pressure sensor, 88, 104, 105, 106, 107

mass air flow sensor, 88, 102, 103

Mediation, 157

Modulator, 40

MOST, 152

Motors, 12

MRE, 87, 92, 93, 95, 96

multiplex, 19

Multiplexed, 92

Negative Temperature Coefficient NTC, 88, 89, 92

network, 153, 155, 157, 158, 160, 161, 162

Networking, 19

neutrons, 10

Node, 152

NOx sensor, 91

Ohm's law, 13

Ohms, 9

open circuit, 15

Open loop, 123

oscilloscope, 21

oxygen sensor, 90, 91, 120, 125, 126

Parallel circuit, 13

parasitic drain, 17

passive sensor, 86

pedal position sensor, 38, 67, 68, 99

periodic table, 10

Personal Protective Equipment (PPE) at Work Regulations 1992, 5

Phase, 129

Photoelectric, 92

physical layer, 154, 157

Piezoelectric, 37, 40

pinking, 90, 118

Pintle, 40

Plenum Chamber, 104

Polarity, 40

Positive Temperature Coefficient PTC, 88, 89, 92

Potential difference, 8, 9, 148, 160

Potentiometer, 92, 191, 192

pre-sets, 32

pressure regulator, 37

Pressure sensors, 172

Pressure transducers, 28

protons, 10

pulse width modulation, 19

Pulse width modulation PWM, 40

Pumpe-Düse, 63

Index

quantity control valve, 38, 65
Rectifier, 40
Relative compression, 73
reluctor, 89, 93, 94, 95, 108, 109, 116
Resistance, 9
RMS, 191
secondary ignition, 25, 131, 141, 144, 149, 186
Sensors, 18
Series circuit, 13
Servo, 40
short circuit, 16
single point fuel injection, 56
Smart charge, 78
Solenoid, 39
Solenoids, 40
starter motor, 39, 71, 72, 73, 74, 75, 182, 193
steering angle sensor, 91, 129, 130
steering torque sensor, 91
stepper motor, 49
Stoichiometric, 125
Stratified, 92
sun load sensor, 91, 128
Sweep, 27
Switches, 86
The Provision and Use of Work Equipment Regulations 1998 (PUWER), 4
Thermistor, 111

Throttle pedal sensors, 98
throttle position sensor, 66, 87
throttle servo, 38
Throttle servo motor, 33, 38, 66, 67
Titania, 90, 120, 121, 124, 125
Tolerance, 92
Topology, 154
Transducer, 26, 172
Transistors, 92
TRC, 87, 92
Triggering, 7
Triggers, 29, 32
Vane air flow meter, 83, 87, 99, 100
Vehicle Protective Equipment (VPE), 5
Vehicle speed sensor, 83, 91, 127
Volts, 8
Volumetric efficiency, 40
vortex air flow sensor, 88, 104
VSC, 87, 92
VVT, 33, 38, 69, 70, 71
wasted spark, 134, 142, 143, 144, 186
Watts, 9
Waveform, 7
Waveforms, 27
wheel speed sensor, 87, 93, 94, 95, 96, 97
Zirconia, 90, 121, 122, 124, 125

If you enjoyed this book, please check out other titles by the author:

Paperback: 338 pages
Publisher: Graham Stoakes
(1 Feb. 2015)
ISBN-10: 099294922X
ISBN-13: 978-0992949228

Level 4 Automotive Master Technician - Advanced Light Vehicle Technology

'Technology needs technicians, and the ability to harness technical diagnosis calls for a Master Technician'.

The rapid growth in technology used in the production of cars has highlighted the need for a different approach to vehicle diagnosis and repair. The integration of complex electronic control with mechanical systems shows the brilliance in the engineering capabilities of designers and manufacturers.

While this technology has improved the comfort, safety, convenience and reliability of vehicles, it has also created an issue with established methods of maintenance and repair. As many of the control systems operate beyond our natural capabilities, diagnostic tooling is required to undertake most of the fault finding duties traditionally conducted by vehicle technicians. Also, the sophisticated nature of advanced system faults will often lead to diagnostic requirements for which there is no prescribed method.

One of the fundamental roles of a Master Technician will be the diagnosis and repair of these complex and advanced system faults, for which diagnostic approaches need to be developed that can provide logical strategies to reduce overall diagnostic time. An effective diagnostic routine should always begin with a logical assessment of symptoms and then uses reasoning to reduce the possible number of options, before following a systematic approach to finding and fixing the root cause.

Paperback: 154 pages
Publisher: Graham Stoakes
(6 July 2015)
ISBN-10: 0992949246
ISBN-13: 978-0992949242

Principles of Light Vehicle Air Conditioning

'As the number of vehicles on the world's roads rises, the demand for increased levels of comfort and convenience also grows'

While air conditioning and climate control may be seen as a luxury by some, the key benefits often outweigh the initial costs and resources required to implement these systems on newly produced vehicles; in fact most new cars come with some form of air conditioning as standard.

An environment which helps keep the driver and passengers comfortable and alert, maintaining the correct levels of ventilation and humidity, can increase concentration and the ability to devote more of their attention to the occupation of driving.

The downside of these systems is the environmental impact of the chemicals used to provide the refrigeration process.

Globally, anthropogenic, or 'man-made' emissions are believed to be the key factor in climate change and refrigerants have a larger influence than many others.

Small amounts of fluorinated gasses released to atmosphere may be causing irreparable damage to our planet, initiating ozone depletion and global warming.

Although many organisations are currently seeking alternatives to these harmful cocktails, at the present time we are restricted by the availability, cost and technology required to make viable replacements.

This means that for the time being, technicians and air conditioning professionals need to ensure that refrigerants are handled with due diligence and systems are maintained to the highest standards in order to contain and reduce emissions. Remember these chemicals only become dangerous when released to atmosphere.

Paperback: 196 pages
Publisher: Graham Stoakes
(1 July 2014)
ISBN-10: 0992949203
ISBN-13: 978-0992949204

Hybrid electric and alternative automotive propulsion

'A future without oil won't spell the end of the car, but will simply drive engineering brilliance to find an alternative'.

As fuel demand and environmental pollution increases, it is important that substitutes are found for traditional methods of vehicle drive. An alternative propulsion vehicle is one that operates using something other than the established petrol or Diesel.

Whether you are a vehicle technician, automotive trainer, student or part of the emergency services, an awareness of current and emerging propulsion sources is vital in order to work on or around these vehicles safely.

www.grahamstoakes.com